It's another Quality Book from CGP

This book is for anyone doing AQA Modular GCSE Mathematics at Foundation Level.

It contains lots of tricky questions designed to make you sweat — because that's the only way you'll get any better.

It's also got some daft bits in to try and make the whole experience at least vaguely entertaining for you.

What CGP is all about

Our sole aim here at CGP is to produce the highest quality books — carefully written, immaculately presented and dangerously close to being funny.

Then we work our socks off to get them out to you — at the cheapest possible prices.

Contents

Unit 1 — Statistics and Number

Unit 2 — Number and Algebra

Unit 3 — Geometry and Algebra

Throughout the book, the more challenging questions are marked like this: **Q1**

Published by Coordination Group Publications Ltd.
Illustrated by Ruso Bradley, Lex Ward and Ashley Tyson

From original material by Richard Parsons.

Updated by:
Simon Little, Ali Palin, David Ryan, Jane Towle and Dawn Wright.

Contributors:
Gill Allen
Margaret Carr
Barbara Coleman
JE Dodds
Mark Haslam
John Lyons
C McLoughlin
Gordon Rutter
John Waller
Janet West
Dave Williams
Philip Wood

Proofreading:
Margaret Darlington and Julie Wakeling.

ISBN: 978 1 84146 557 9

Groovy website: www.cgpbooks.co.uk

Printed by Elanders Hindson Ltd, Newcastle upon Tyne.
Clipart sources: CorelDRAW® and VECTOR.

Rounding Off

Q1 Round the following to the nearest whole number:

a) 2.9 .... **b)** 26.8 .... **c)** 2.24 ....

d) 11.11 .... **e)** 6.347 .... **f)** 43.5 ....

g) 9.99 .... **h)** 0.41 ....

Nearest whole number means you look at the digit after the decimal point to decide whether to round up or down.

Q2 An average family has 2.3 children, how many children is this to the nearest whole number?

....

Q3 By the time she is 25 the average woman will have driven 4.72 cars. What is this to the nearest whole number?

....

Q4 Give these amounts to the nearest pound:

a) £4.29 .... **b)** £16.78 .... **c)** £12.06 ....

d) £7.52 .... **e)** £0.93 .... **f)** £14.50 ....

g) £7.49 .... **h)** £0.28 ....

Q5 Give these amounts to the nearest hour:

a) 2 hours 12 minutes .... **b)** 36 minutes ....

c) 12 hours 12 minutes .... **d)** 29 minutes ....

e) 100 minutes .... **f)** 90 minutes ....

Rounding Off

Q6 Round off these numbers to the nearest 10:

a) 23 .... **b)** 78 .... **c)** 65 .... **d)** 99 ....

e) 118 .... **f)** 243 .... **g)** 958 .... **h)** 1056 ....

Q7 Round off these numbers to the nearest 100:

a) 627 .... **b)** 791 .... **c)** 199 .... **d)** 450 ....

e) 1288 .... **f)** 3329 .... **g)** 2993 ....

Q8 Round these off to the nearest 1000:

a) 5200 .... **b)** 8860 .... **c)** 9870 ....

Q9 Crowd sizes at sports events are often given exactly in newspapers. Round off these exact crowd sizes to the nearest 1000:

a) 23 324 ....

b) 36 844 ....

c) 49 752 ....

Q10 The number of drawing pins in the box has been rounded to the nearest 10.

DRAWING PINS
Contents: 80

What is the least possible number of drawing pins in the box? ....

What is the greatest possible number? ....

Q11 The population of Whichtown is given as 1300 to the nearest 100.
What is the smallest number the population could be? ....

What is the largest it could be? ....

When you round numbers off to the nearest unit, the ACTUAL measurement could be up to HALF A UNIT bigger or smaller...

Rounding Off

Q12 Round off these numbers to 1 decimal place (1 d.p.):

a) 7.34 .... **b)** 8.47 .... **c)** 12.08 .... **d)** 28.03 ....

e) 9.35 .... **f)** 14.618 .... **g)** 30.409 ....

Q13 Round off the following to 2 d.p.:

a) 17.363 .... **b)** 38.057 .... **c)** 0.735 ....

d) 5.99823 .... **e)** 4.297 .... **f)** 7.0409 ....

Q14 Now round these to 3 d.p.:

a) 6.3534 .... **b)** 81.64471 .... **c)** 0.0075 ....

d) 53.26981 .... **e)** 754.39962 .... **f)** 0.000486 ....

Q15 Seven people have a meal in a restaurant. The total bill comes to £60. If they share the bill equally, how much should each of them pay?
Round your answer to 2 d.p.

....

Q16 Round these numbers to 1 significant figure.

a) 12 .... **b)** 530 ....

c) 1379 .... **d)** 0.021 ....

e) 1829.62 .... **f)** 0.296 ....

The <u>1st significant figure</u> of any number is the first digit which isn't zero.

Q17 Matt is buying skirting board to fit to a wall which measures 540 cm to 2 s.f. What length of board should Matt buy to guarantee he has enough for the whole wall?

....

Calculator Questions

Yeah, OK, we all know how to do sums on a calculator — but it can do so much more... check out the groovy powers button and the funky brackets buttons, not to mention the slinky a b/c button...

Q1 Using the x^2 button on your calculator, work out:

a) 1^2 .... **c)** 16^2 .... **e)** $(-5)^2$ ....

b) 2^2 .... **d)** $(-1)^2$ .... **f)** 1000^2 ....

Q2 Using the x^y or ^ button, find:

a) 1^3 .... **c)** 4^5 .... **e)** 2^{-1} ....

b) 3^6 .... **d)** 3^{-2} .... **f)** 0^2 ....

Q3 Using the √ button on your calculator, work out:

a) $\sqrt{16}$ .... **d)** $\sqrt{0}$ .... **g)** $\sqrt{3}$ ....

b) $\sqrt{36}$ .... **e)** $\sqrt{3600}$ .... **h)** $\sqrt{7}$ ....

c) $\sqrt{289}$ .... **f)** $\sqrt{400}$ .... **i)** $\sqrt{30}$ ....

Q4 Using (and), calculate:

a) $\frac{(14+18)}{(2\times 8)}$ .... b) $\frac{(9+(4\div 2))}{(11\times 3)}$ .... c) $\frac{12}{(8+9)(13-11)}$ ....

Q5 Using the a b/c button on your calculator, work out:

a) $\frac{2}{5}+\frac{3}{4}$ .... **d)** $\frac{2}{5}\times\frac{3}{4}$ .... **g)** $\frac{1}{7}\times\frac{1}{8}\times\frac{1}{9}$ ....

b) $\frac{1}{8}-\frac{1}{9}$ .... **e)** $\frac{7}{8}\times\frac{1}{16}$ .... **h)** $\frac{350}{490}$ (lowest terms)

c) $\frac{4}{13}-\frac{1}{8}+\frac{2}{7}$ .... **f)** $\frac{5}{12}\div\frac{4}{11}$ ....

Q6 Use your calculator to work out:

a) the cost of 5 sherbet beetles for 56p each and 10 fizzy slugs for £1.21 each. ....

b) the cost to the nearest penny of 1 cola emu if a pack of 36 costs £2.49. ....

Fractions

This circle is divided into two equal parts — each bit is 1/2 of the whole.

Q1 What fraction is shaded in each of the following pictures?

a) 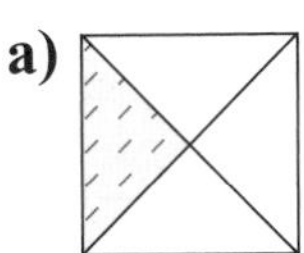

....

b) 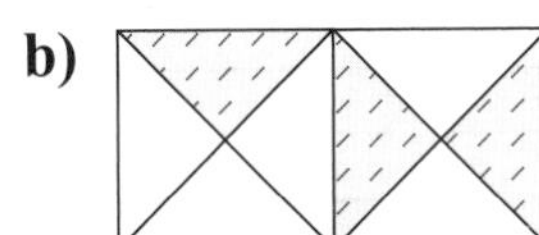

....

c) 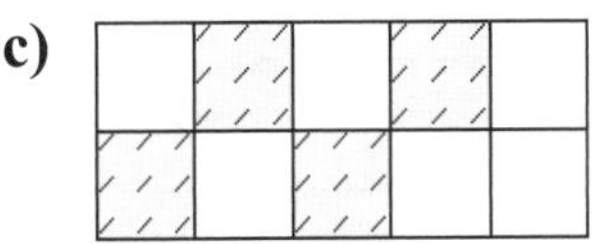

....

d) 

....

e) 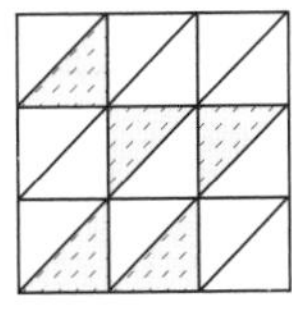

....

f) 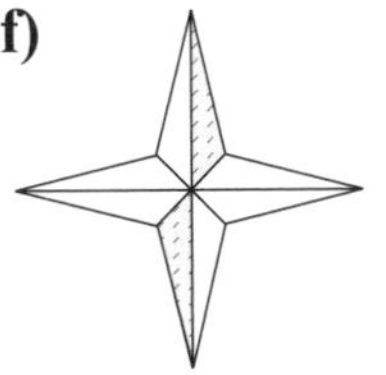

....

g) 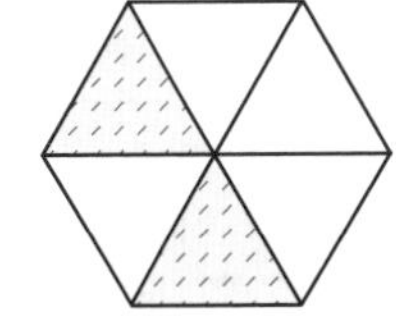

....

Q2 Shade the diagram to show the following fractions.

a) $\frac{3}{8}$

b) $\frac{2}{5}$

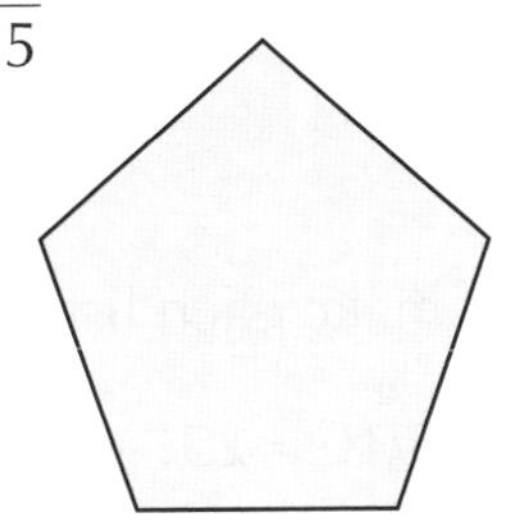

c) $\frac{7}{10}$

Q3 There are 32 boxes of shoes stacked on a shelf.

If five eighths are sold, show on the diagram how many are left.

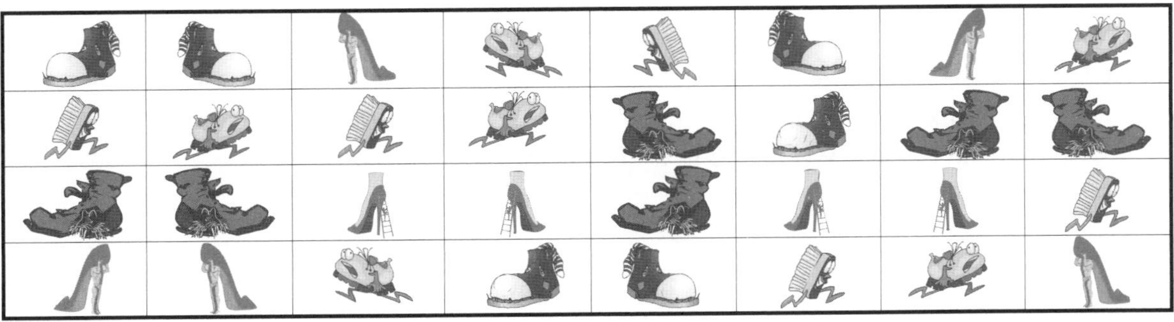

Fractions

To make an EQUIVALENT fraction, you've got to multiply the TOP (numerator) and BOTTOM (denominator) by the SAME THING.

Q4 Shade in the correct number of sections to make these diagrams equivalent.

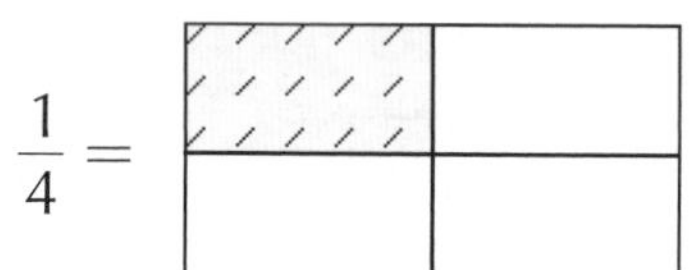

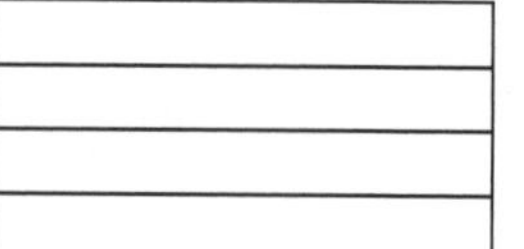
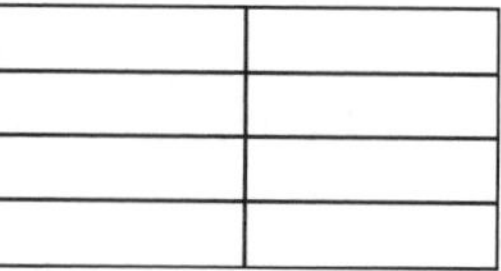
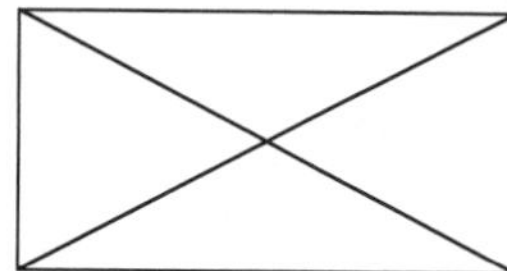

$\frac{1}{3}$ =

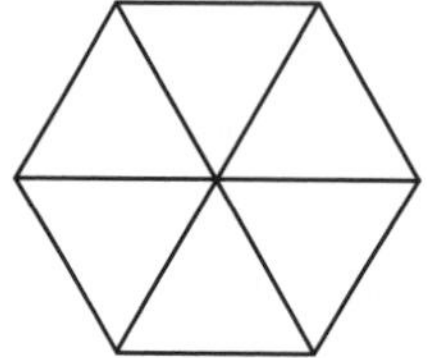
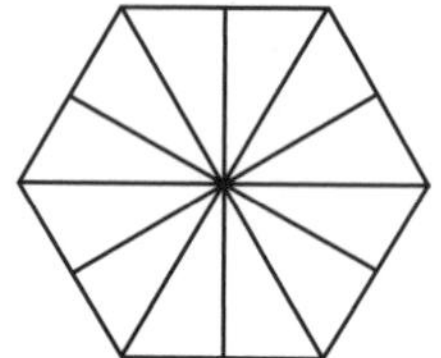
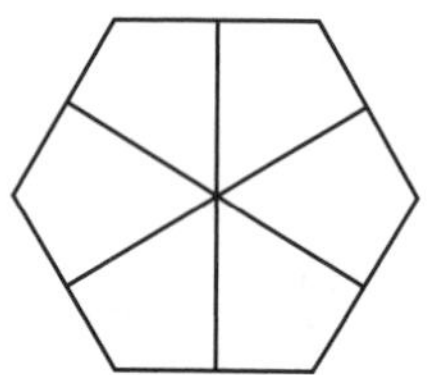

Q5 Write in the missing numbers to make these fractions equivalent.

For example 1/2 = 7/14

a) 1/4 = 4/........ **b)** 3/4 = 9/........ **c)** 1/3 = ... /6

d) 2/3 = 8/........ **e)** 6/18 = 1/........ **f)** 24/32 = 3/........

Q6 Write in the missing numbers to make each list equivalent.

a) 1/2 = 2/...... = ... /6 = ... /8 = 5/10 = 25/...... = ... /70 = ... /100

b) 200/300 = 100/ ...... = ... /15 = 40/ ...... = 120/180 = ... /9 = ... /3

c) 7/10 = 14/ ...... = ... /30 = 210/ ...... = 49/ ...... = ... /20

d) 19/20 = ... /80 = 38/ ...... = 57/ ...... = ... /100 = ... /1000

Q7 Healthybix cereal contains 15 g of sugar in every 100 g.
How much sugar would be in a 60 g bowl of Healthybix?

....

Fractions of Quantities

Q1 Write down the following quantities:

a) Half of 12 = ... **b)** Quarter of 24 = ... **c)** Third of 30 = ...

d) 1/4 of 44 = ... **e)** 3/4 of 60 = ... **f)** 2/3 of 6 = ...

Q2 Calculate the fractions of the following:

eg. 1/3 of 18 = 18 ÷ 3 = <u>6</u>

a) 1/8 of 32 = ... ÷ 8 = ...

b) 1/10 of 50 = ... ÷ 10 = ...

c) 1/12 of 144 = ... ÷ .... = ...

d) 1/25 of 75 = ... ÷ .... = ...

e) 1/30 of 180 = ... ÷ .... = ...

f) 1/27 of 540 = ... ÷ .... = ...

Q3 Calculate the fractions of the following:

eg. 2/5 of 50 50 ÷ 5 = 10 2 × 10 = <u>20</u>

a) 2/3 of 60 60 ÷ 3 = ... 2 × .... = ...

b) 4/5 of 25 25 ÷ .... = ... 4 × .... = ...

c) 7/9 of 63 .... ÷ .... = ...× .... = ...

d) 3/10 of 100 .... ÷ .... = ... × .... = ...

e) 12/19 of 760 .... ÷ .... = ... × .... = ...

f) 6/9 of £1.80 .... ÷ .... = ... × .... = £........ or .... p

g) 10/18 of £9.00 .... ÷ .... = ... × .... = £........

Q4 A return car journey from Lancaster to Stoke uses 5/6 of a tank of petrol. How much does this cost, if it costs £54 for a full tank of petrol?

....

Q5 Peter works in a clothes shop. He gets a staff discount and only has to pay 2/3 of the price of any item. How much will a £24 top cost him?

....

Q6 A recipe for 5 people needs 1 kg of potatoes. Helen reduces the recipe to make enough for 2 people. How many kilograms of potatoes does she need?

....

Fraction Problems

Q1 What fraction of 1 hour is:

a) 5 minutes? ....

b) 15 minutes? ....

c) 40 minutes? ....

Q2 If a TV programme lasts 40 minutes, what fraction of the programme is left after:

a) 10 minutes? ....

b) 15 minutes? ....

c) 35 minutes? ....

Q3 A café employs eighteen girls and twelve boys to wait at tables. Another six boys and nine girls work in the kitchen.
What fraction of the kitchen staff are girls?
What fraction of the employees are boys?

Fraction of kitchen staff who are girls ....

Fraction of employees who are boys ....

Q4 At a college, two fifths of students are female, and three sevenths of these are part-time. If there are 2100 students altogether, how many:

a) are female students? ....

b) are part-time female students? ....

Q5 If I pay my gas bill within seven days, I get a reduction of an eighth of the price. If my bill is £120, how much can I save? ....

Q6 Owen gives money to a charity directly from his wages, before tax. One sixth of his monthly earnings goes to the charity, then one fifth of what's left gets deducted as tax. How much money goes to charity and on tax, and how much is left, if he earns £2400?

Charity ...., Tax ...., Left over ....

Q7 In the summer three friends ran a car-cleaning service. They divided up the profits at the end of the summer according to the proportion of cars each had cleaned. Ali had washed 200 cars, Brenda had washed 50 and Chay had washed 175. The profits were £1700. How much did each person get?

Ali ...., Brenda ...., Chay ....

This is similar to Q6, except you've got to work out the fractions of the money they get yourself.

Don't get put off by all the padding in the questions — you've just got to pick out the important stuff.

Percentages

Finding "something %" of "something-else" is really quite simple — so you'd better be sure you know how.

1) "OF" means "×". 2) "PER CENT" means "OUT OF 100".	Example: 30% of 150 would be translated as $\frac{30}{100} \times 150 = 45$.

Q1 Try these without a calculator:

a) 50% of £12 = ... **b)** 25% of £20 = ...

c) 10% of £50 .... **d)** 5% of £50 = ...

e) 30% of £50 = ... **f)** 75% of £80 = ...

g) 10% of 90 cm = ... **h)** 10% of 4.39 kg = ...

Now you can use a calculator:

i) 8% of £16 = ...

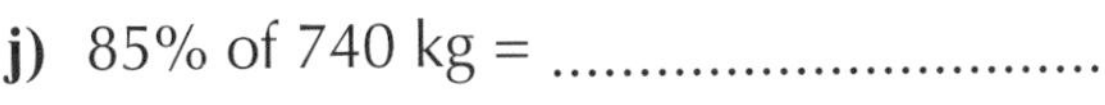

j) 85% of 740 kg = ...

k) 40% of 40 minutes = ...

A school has 750 pupils.

l) If 56% of the pupils are boys, what percentage are girls? ....

m) How many boys are there in the school? ....

n) One day, 6% of the pupils were absent. How many pupils was this? ....

o) 54% of the pupils have a school lunch, 38% bring sandwiches and the rest go home for lunch. How many pupils go home for lunch?

....

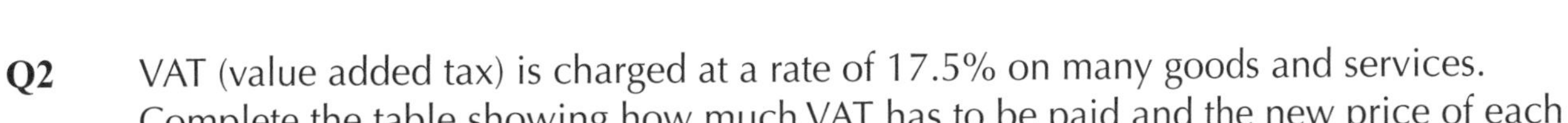

Q2 VAT (value added tax) is charged at a rate of 17.5% on many goods and services. Complete the table showing how much VAT has to be paid and the new price of each article.

Article	Basic price	VAT at 17.5%	Price + VAT
Tin of paint	£6.75		
Paint brush	£3.60		
Sand paper	£1.55		

Percentages

Q3 John bought a new PC. The tag in the shop said it cost £890 + VAT.
If VAT is charged at 17½%, how much did he pay?

....

Q4 Admission to Wonder World is £18 for adults. A child ticket is 60% of the adult price.

a) How much will it cost for one adult and 4 children to enter Wonder World?

....

b) How much will two adults and three children spend on entrance tickets?

....

Q5 Daphne earns an annual wage of £18 900. She doesn't pay tax on the first £6400 that she earns. How much income tax does she pay if the rate of tax is:

a) 20% ? .... **b)** 40% ? ....

Q6 Terence paid £4700 for his new motorcycle. Each year its value decreased by 12%.

a) How much was it worth when it was exactly one year old? ....

b) How much was it worth when it was exactly two years old? ....

Q7 A double-glazing salesman is paid 10% commission on every sale he makes.
In addition he is paid a £50 bonus if the sale is over £500.

a) If a customer buys £499 worth of windows from the salesman, what is his commission?

....

b) How much extra will he earn if he persuades the customer in **a)** to spend an extra £20?

....

c) How much does he earn if he sells £820 worth of windows?

....

Q8 **Bed & breakfast £37 per person. Evening meal £15 per person.**

2 people stay at The Pickled Parrot for 2 nights and have an evening meal each night. How much is the total cost, if VAT is added at 17½% ?

....

THE PICKLED PARROT

Percentages

Divide the new amount by the old, then × 100... or if you've been practising on your calc, you'll know you can just press the % button for the 2nd bit...

Q9 Express each of the following as a percentage. Round off if necessary.

a) £6 of £12 = ...

b) £4 of £16 = ...

c) 600 kg of 750 kg = ...

d) 6 hours of one day = ...

e) 1 month of a year = ...

f) 25 m of 65 m = ...

Q10 Calculate the percentage saving of the following:

e.g. Trainers: Was £72 Now £56 Saved ..£16.. of £72 = ...22....%

a) Jeans: Was £45 Now £35 Saved .... of £45 = ...%

b) CD: Was £14.99 Now £12.99 Saved .... of .... = ...%

c) Shirt: Was £27.50 Now £22.75 Saved .... of .... = ...%

d) TV: Was £695 Now £435 Saved .... = ...%

e) Microwave: Was £132 Now £99 Saved .... = ...%

Q11 a) If Tim Hangman won 3 out of the 5 sets in the Wimbledon Men's Final, what percentage of the sets did he not win?

....%

b) Of a resort's 25 000 tourists last Summer, 23 750 were between 16 and 30 years old. What percentage of the tourists were not in this age group?

....%

c) Jeff went on a diet. At the start he weighed 92 kg, after one month he weighed 84 kg. What is his percentage weight loss?

....%

d) In their first game of the season, Nowcastle had 24 567 fans watching the game. By the final game there were 32 741 fans watching. What is the percentage increase in the number of fans?

....%

Ratios

Ratios compare quantities of the same kind — so if the units aren't mentioned, they've got to be the same in each bit of the ratio.

Q1 What is the ratio in each of these pictures?

a)

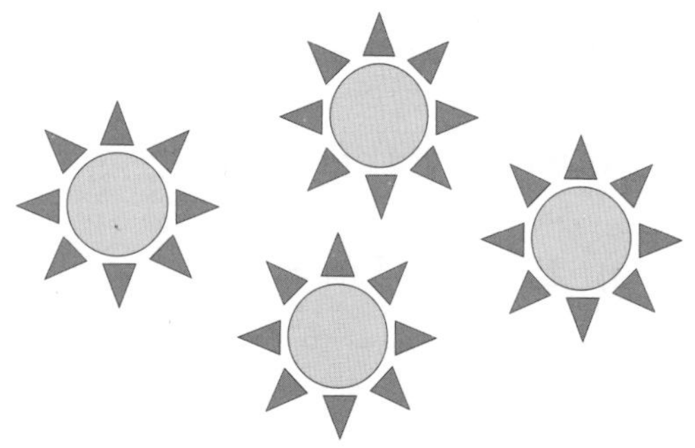

Circles to triangles

.... to ....

Triangles to circles

.... : ....

b)

Small stars to big stars

.... to ....

Big stars to small stars

.... : ....

Q2 In 'The Pink Palace' the walls are painted a delicate shade of pink using 2 tins of red paint to every 3 tins of white paint.

Write this as a ratio.

Red tins .... to White tins ....

R........... : W...........

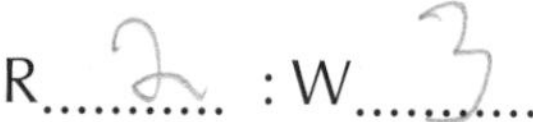

Q3 Write each of these ratios in its simplest form. The first one is done for you.

a) 4 to 6

2 : 3

b) 15 to 21

...... : ......

c) 14 to 42

...... : ......

d) 72 to 45

...... : ......

e) 24 cm to 36 cm

...... : ......

f) 350 g to 2 kg

...... : ......

g) 42p to £1.36

...... : ......

Watch out for ones like f) and g) — you need to make the units the same first.

Ratios

I think I can spot a Golden Rule lurking here...
DIVIDE FOR ONE, THEN TIMES FOR ALL.

Q4 To make grey paint, black and white paint are mixed in the ratio 5:3.
How much black paint would be needed with:

a) 6 litres of white? 10 l

b) 12 litres of white? 20 l

c) 21 litres of white? 35 l

Q5 To make salad dressing you mix olive oil and vinegar in the ratio 5:2.
How much olive oil is needed with:

a) 10 ml of vinegar? 25 ml

b) 30 ml of vinegar? 75 ml

c) 42 ml of vinegar? 105

Q6 Sarah works as a waitress. Each week, she splits her wage into spending money and savings in the ratio 7:3.

a) One week, Sarah has £91 spending money.
How much should she put in her savings that week? £39

b) The next week, Sarah put £42 into her savings.
How much did she earn in total that week? £140

Q7 The ratio of men to women at a football match was 11:4.
How many men were there if there were:

a) 2000 women? 5500

b) 8460 women? ....

Best Buys

Start by finding the AMOUNT PER PENNY — the more of the stuff you get per penny, the better value it is.

Q1

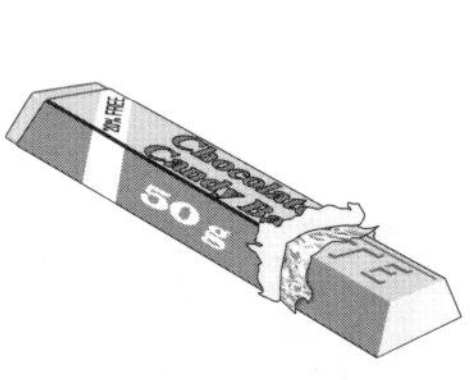

The small bar of chocolate weighs 50 g and costs 32p.
The large bar weighs 200 g and costs 80p.

a) How many grams do you get for 1p from the small bar?

....

b) How many grams do you get for 1p from the large bar?

....

c) Which bar gives you more for your money?

....

Q2 The large tin of tuna weighs 400 g and costs £1.05.
The small tin weighs 220 g and costs 57p.

a) How many grams do you get for 1p in the large tin?

....

b) How many grams do you get for 1p in the small tin?

....

c) Which tin gives better value for money?

....

Q3 Which of these boxes of eggs is better value for money?

60p

£1.10

....

....

Collecting Data

Q1 Zac collects some data about his school. The data items are listed below. Say whether each data item is qualitative or quantitative.

a) The colours of pants worn by the teachers. ....

b) The distance travelled to school by each student. ....

c) The star sign of each student. ....

Q2 Complete the table to show which type of data has been used.

Data	Discrete or Continuous
Number of people passing through Heathrow airport each day.	
Heights of people in a Maths class.	
Number of goals scored in a hockey game.	
Times taken to be served in a fast-food restaurant.	

Don't forget: if you can measure a quantity exactly then it's discrete — otherwise it's continuous.

Q3 Stanley is researching the use of the school canteen.
He asks this question to a sample of students at the school:
How often do you use the canteen? Tick one of the boxes.

Very often ☐ *Quite often* ☐

Not very often ☐ *Never* ☐

Questions can be either Open or Closed — make sure you know what this means.

a) Draw a tally table that Stanley could use as an observation sheet to collect his data.

b) Give one criticism of Stanley's question.

....

c) Write a better question that Stanley could use to find out how often students at his school use the canteen.

....

Collecting Data

If you're doing stats it's easy to have loads and loads of data, which can make it pretty hard to analyse. So it helps to group it together so it's more manageable.

Q4 Fred asked each of his 30 classmates how long (in minutes) it took them to eat their dinner. Here are the results he recorded:

42	13	3	31	15	20	19	1	59	14
8	25	16	27	4	55	32	31	31	10
32	17	16	19	29	42	43	30	29	18

a) Group the data appropriately and fill in the table.

Length of time (mins)					
Number of people					

b) State one disadvantage of grouping data.

....

Q5 Millom University Chemistry Department wants to find out the influence of its marketing on sixth-form chemistry students in the UK. They compile a survey and send it to all the students at the nearest sixth-form college.

a) Give one reason why this sample is biased.

....

b) What population should the Millom University Chemistry Department be sampling from?

....

Q6 The Cheapeez discount food chain wants to find out what products to stock to attract more of the residents of Devon. Their interviewers survey the first 200 people to go into five Cheapeez supermarkets in Devon on the first Saturday in March.

a) Give one reason why this sample will produce biased data.

....

b) What population should have been sampled from?

....

Mode and Median

To find the mode, put the data in order of size first — then it's easier to see which number you've got most of.

Q1 Find the mode for each of these sets of data.

a) 3, 5, 8, 6, 3, 7, 3, 5, 3, 9

.... Mode is ....

b) 52, 26, 13, 52, 31, 12, 26, 13, 52, 87, 41

.... Mode is ....

Q2 The temperature in °C on 10 summer days in England was:

25, 18, 23, 19, 23, 24, 23, 18, 20, 19

What was the modal temperature?

.... Modal temperature is ....°C.

Q3 The time it takes thirty pupils in a class to get to school each day in minutes is:

18, 24, 12, 28, 17, 34, 17, 17, 28, 12, 23, 24, 17, 34, 9,
32, 15, 31, 17, 19, 17, 32, 15, 17, 21, 29, 34, 17, 12, 17

What is the modal time?

....

....

Modal time is .... mins.

Put the data in order of size for median questions too — it's much easier to find the middle value.

Q4 Find the median for these sets of data.

a) 3, 6, 7, 12, 2, 5, 4, 2, 9

.... Median is ....

b) 14, 5, 21, 7, 19, 3, 12, 2, 5

.... Median is ....

Q5 These are the heights of fifteen 16 year olds.

162 cm 156 cm 174 cm 148 cm 152 cm 139 cm 167 cm 134 cm
157 cm 163 cm 149 cm 134 cm 158 cm 172 cm 146 cm

What is the median height? Write your answer in the shaded box.

															Median

Mean and Range

Yikes — MEAN questions... well, they're not as bad as everyone makes out. Remember to include zeros in your calculations — they still count.

Q1 Find the mean of each of the sets of data below. If necessary, round your answers to 1 decimal place:

Remember the formula for mean — total of the items ÷ number of items.

a) 13, 15, 11, 12, 16, 13, 11, 9 = ...

b) 16, 13, 2, 15, 0, 9 = ...

c) 80, 70, 80, 50, 60, 70, 90, 60, 50, 70, 70 = ...

Q2 Find the mean for each of these sets of data:

a) 5, 3, 7, 3, 2 = ...

b) 7, 3, 9, 5, 3, 5, 4, 6, 2, 6 = ...

Q3 David is on a diet. He needs to have a mean calorie intake of 1900 calories a day. The table below shows how many calories he has eaten so far this week.

How many calories must David eat on Friday to make sure that his mean intake for the five days is correct?

....

Day	Calories Eaten
Monday	1875
Tuesday	2105
Wednesday	1680
Thursday	1910
Friday	

Q4 The number of goals scored by a hockey team over a period of 10 games is listed below.

0, 3, 2, 4, 1, 2, 3, 4, 1, 0.

What is the range of the number of goals scored? ....

Q5 Sarah and her friends were measured and their heights were found to be:

1.52 m, 1.61 m, 1.49 m, 1.55 m, 1.39 m, 1.56 m.

What is the range of the heights? ....

Q6 Here are the times 6 people took to do a Maths test:

1 hour 10 mins, 2 hours 10 mins, 1 hour 35 mins,
1 hours 55 min, 1 hour 18 mins, 2 hours 15 mins.

What is the range of these times? ....

Averages

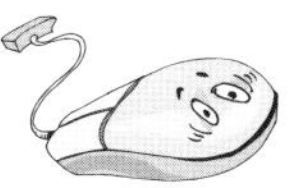

Geesh — as if it's not enough to make you work out all these boring averages, they want you to write stuff about them as well. Oh well, here goes nothing.

Q1 These are the mathematics marks for John and Mark.

John	65	83	58	79	75
Mark	72	70	81	67	70

Calculate the mean and range for each pupil. Who do you think is the better maths student? Why?

....

....

Q2 Jane has a Saturday job. She earns £3.70 an hour. She thinks that most of her friends earn more. Here is a list of how much an hour her friends are paid.

Lisa £3.60 Scott £4.50 Kate £4.25 Kylie £3.40
Helen £4.01 Kirsty £4.25 Ben £4.25 Ruksana £3.90

Work out the mean, median and mode for her friends' pay.

....

....

Jane's employer offers her a rise to £4.02 because she claims that this is the average hourly rate. Which average has her employer used?

....

Which average should Jane use to try to negotiate a higher pay rise?

....

Q3 The number of absences for 20 pupils during the spring term were:

0 0 0 0 0 0 1 1 1 2 3 4 4 4 7 9 10 10 19 22

Work out the mean, median and modal number of absences.

....

....

If you were a local newspaper reporter wishing to show that the local school has a very poor attendance record, which average would you use and why?

....

If you were the headteacher writing a report for the parents of new pupils, which average would you use and why?

....

Averages

Q4 On a large box of matches it says "Average contents 240 matches".
I counted the number of matches in ten boxes. These are the results:

241 244 236 240 239 242 237 239 239 236

Is the label on the box correct? Use the mean, median and mode for the numbers of matches to explain your answer.

....

....

Q5 The shoe sizes in a class of girls are:

3 3 4 4 5 5 5 5 6 6 6 7 8

Calculate the mean, median and mode for the shoe sizes.

....

....

If you were a shoe shop manager, which average would be most useful to you, and why?

....

Q6 A house-building company needs a bricklayer.
This advert appears in the local newspaper.
The company employs the following people:

Position	Wage
Director	£700
Foreman	£360
Plasterer	£300
Bricklayer	£250
Bricklayer	£250

Bricklayer wanted
Average wage over £350 p.w.

What is the median wage?

....

What is the mean wage? ....

Which gives the best idea of the average wage? ....

Is the advert fair? Explain your answer.

....

....

Write a fairer advert in the space above.

Frequency Tables

Q1 At the British Motor Show 60 people were asked what type of car they preferred. Jeremy wrote down their replies using a simple letter code.

Saloon - S Hatchback - H 4x4 - F MPV - M Roadster - R

Here is the full list of replies.

H	S	R	S	S	R	M	F	S	S	R	R
M	H	S	H	R	H	M	S	F	S	M	S
R	R	H	H	H	S	M	S	S	R	H	H
H	H	R	R	S	S	M	M	R	H	M	H
H	S	R	F	F	R	F	S	M	S	H	F

Fill in the tally table and add up the frequency in each row. Draw the frequency graph of the results.

A frequency graph just means a bar chart.

TYPE OF CAR	TALLY	FREQUENCY
Saloon		
Hatchback		
4 × 4		
MPV		
Roadster		

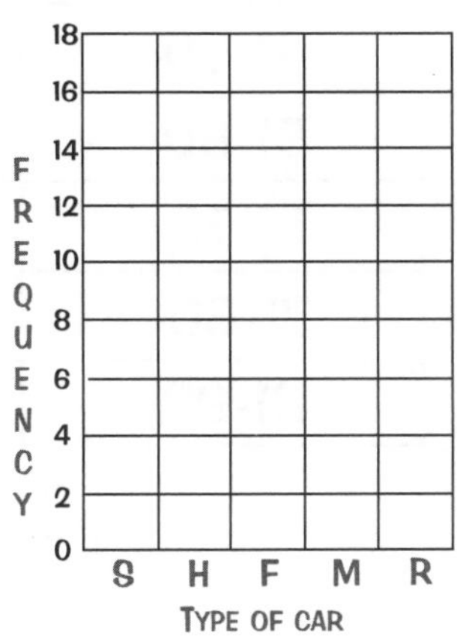

Q2 Last season Newcaster City played 32 matches.
The number of goals they scored in each match was recorded as shown.

2	4	3	5
1	0	3	2
4	2	1	2
0	2	3	1

1	0	0	1
1	1	1	0
1	3	2	0
1	1	0	4

Complete the tally chart and draw the frequency polygon of the goals.

GOALS	TALLY	FREQUENCY
0		
1		
2		
3		
4		
5		

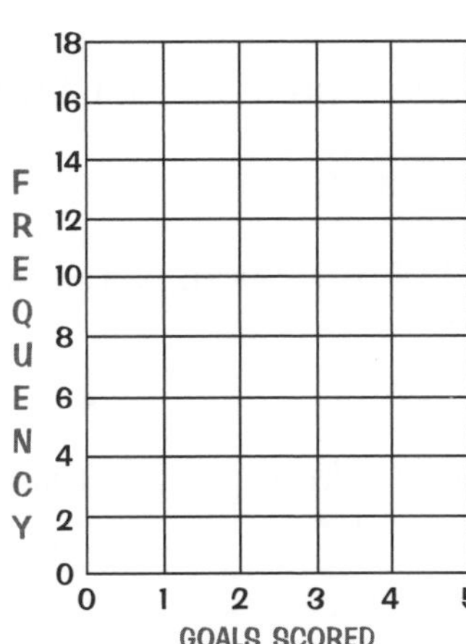

A frequency polygon is where you plot points for the frequencies and join them up with straight lines.

Frequency Tables

Q3 Here is a list of marks which 32 pupils gained in a History test:

65	78	49	72	38	59	63	44
55	50	60	73	66	54	42	72
33	52	45	63	65	51	70	68
84	61	42	58	54	64	75	63

Complete the tally table making sure you put each mark in the correct group
Then fill in the frequency column.

MARKS	TALLY	FREQUENCY
31-40		
41-50		
51-60		
61-70		
71-80		
81-90		
	TOTAL	

Q4 The frequency table below shows the number of hours spent Christmas shopping by 100 people surveyed in a town centre.

Number of Hours	0	1	2	3	4	5	6	7	8
Frequency	1	9	10	10	11	27	9	15	8
Hours × Frequency									

a) What is the modal number of hours spent Christmas shopping? ....

b) Fill in the third row of the table.

c) What is the total amount of time spent Christmas shopping by all the people surveyed?

....

d) What is the mean amount of time spent Christmas shopping by a person?

....

Grouped Frequency Tables

As a rule these are trickier than standard frequency tables — you'll certainly have to tread carefully here. Have a good look at the box below and make sure you remember it.

Mean and Mid-Interval Values

1) The MID-INTERVAL VALUES are just what they sound like — the middle of the group.
2) Using the Frequencies and Mid-Interval Values you can estimate the MEAN.

$$\text{Estimated Mean} = \frac{\text{Overall Total (Frequency} \times \text{Mid-interval value)}}{\text{Frequency total}}$$

Shoe Size	1 - 2	3 - 4	5 - 6	7 - 8	Totals
Frequency	15	10	3	1	29
Mid-Interval Value	1.5	3.5	5.5	7.5	—
Frequency x Mid-Interval Value	22.5	35	16.5	7.5	81.5

So the estimated mean value is 81.5 ÷ 29 = 2.81

Q1 In a survey of test results in a French class at Blugdon High, these grades were achieved by the 23 pupils:

(grade) score	(E) 31-40	(D) 41-50	(C) 51-60	(B) 61-70
frequency	4	7	8	4

a) Write down the mid-interval values for each of the groups.

....

b) Calculate an estimate for the mean value.

....

....

Q2 Dean is carrying out a survey for his Geography coursework. He asked 80 people how many miles they drove their car last year to the nearest thousand. He has started filling in a grouped frequency table to show his results.

No. of Miles (thousands)	1 - 10	11 - 20	21 - 30	31 - 40	41 - 50	51 - 60	61 - 70	71 - 80	81 - 90	91 - 100
No. of Cars	2	3	5	19	16	14	10	7		

a) Complete Dean's table using the following information:

81 245, 82 675, 90 159, 90 569

b) Write down the modal class. ....

c) Which class contains the median number of miles driven? ....

Grouped Frequency Tables

Q3 This table shows times for each team of swimmers, the Dolphins and the Sharks.

Dolphins			Sharks		
Time interval (seconds)	Frequency	Mid-interval value	Time interval (seconds)	Frequency	Mid-interval value
$14 \le t < 20$	3	17	$14 \le t < 20$	6	17
$20 \le t < 26$	7	23	$20 \le t < 26$	15	23
$26 \le t < 32$	15		$26 \le t < 32$	33	
$32 \le t < 38$	32		$32 \le t < 38$	59	
$38 \le t < 44$	45		$38 \le t < 44$	20	
$44 \le t < 50$	30		$44 \le t < 50$	8	
$50 \le t < 56$	5		$50 \le t < 56$	2	

a) Complete the table, writing in all mid-interval values.

b) Use the mid-interval technique to estimate the mean time for each team.

....

....

Q4 The lengths of 25 snakes are measured to the nearest cm, and then grouped in a frequency table.

Length	151 - 155	156 - 160	161 -165	166 - 170	171 - 175	Total
frequency	4	8	7	5	1	25

Which of the following sentences may be true and which have to be false?

a) The median length is 161 cm. ....

b) The range is 20 cm. ....

c) The modal class has 7 snakes.

....

d) The median length is 158 cm.

....

Remember — with grouped data you can't find these values exactly, but you can narrow down the possibilities.

Line Graphs

Drawing line graphs is easy — just plot the points, then join them up with straight lines.

Q1 Billy took his temperature and recorded it on this graph.

What was his temperature at:

a) 10 am? ....

b) 2 pm? ....

c) What was his highest temperature?

....

d) When was this? ....

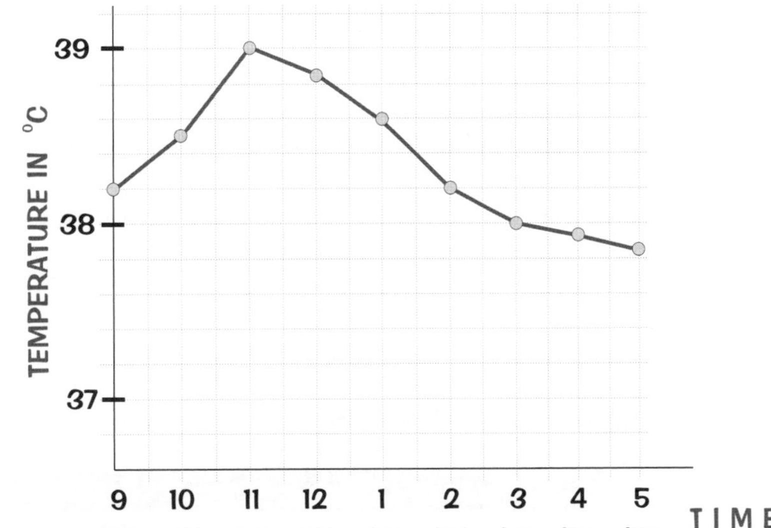

Q2 David weighed himself every 5 days. The results are given below. Draw a graph to show how David's weight changed.

DAY N°	0	5	10	15	20	25	30
WEIGHT KG	88.9	88.4	87.6	87.5	87.5	87.2	87.0

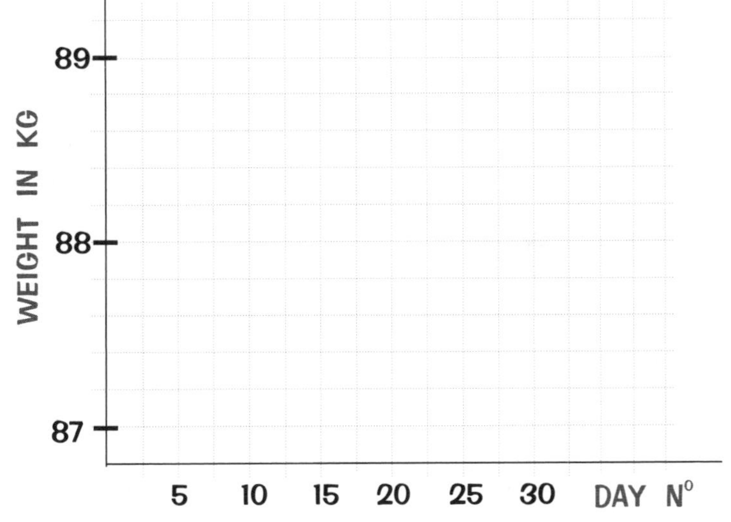

In your own words describe how David's weight changed:

....

....

....

....

Tables, Charts and Graphs

Make sure you read these questions carefully. You don't want to lose easy marks by looking at the wrong bit of the table or chart.

Q1 One hundred vehicles on a road were recorded as part of a traffic study. Use this two-way table to answer the following questions.

	Van	Motor-bike	Car	Total
Travelling North	15			48
Travelling South	20		23	
Total		21		100

a) How many vans were recorded? ....

b) How many vehicles in the survey were travelling south? ....

c) How many motorbikes were travelling south? ....

d) How many cars were travelling north? ....

Q2 This pictogram shows the favourite drinks of a group of pupils.

Favourite Drinks	Number of Pupils
Lemonade	✧ ✧ ✧ ✧ ✧ ✧ ✧ ✧ ✧
Cola	✧ ✧ ✧ ✧ ✧ ✧ ✧ ✧ ✧ ✧ ✧
Cherryade	✧ ✧ ✧ ✧ ✧ ✧
Orange Squash	✧ ✧ ✧
Milk	✧

✧ Represents 2 pupils.

a) How many pupils were questioned? .... pupils.

b) How many pupils prefer non-fizzy drinks? .... pupils.

c) 18 pupils liked lemonade best. How many more liked cola best? .... pupils.

d) Comment on the popularity of cola compared with milk.

....

Q3 Carol was asked to find out if most people have a calculator in their Maths lessons. She asked the people she liked in her Maths set if they had brought a calculator to school. This table shows their responses:

Yes	No
||	|||

From this she claims most people do not have a calculator.
Give 3 criticisms of Carol's survey.

....

....

Tables, Charts and Graphs

Q4 This stem and leaf diagram shows the ages of people in a cinema.

Stem	Leaf
1	2 2 4 8 8 9 9
2	0 1 1 2 5 6
3	0 0 0 5
4	2 5 9
5	
6	8

Key: 2 | 5 means 25

a) How many people in the cinema were in their twenties? ....

b) Write out the ages of all of the people in the cinema below.

....

Q5 This stem and leaf diagram shows the exam scores of a group of Year 9 pupils.

a) How many pupils got a score between 60 and 70? ....

b) How many scored 80 or more? ....

c) What was the most common test score? ....

d) How many scored less than 50? ....

e) How many pupils took the test? ....

f) What was the median test score? ....

g) What was the range of the test scores? ....

Stem	Leaf
3	2 3
4	6 8 8
5	1 2 2 3 6 6 9
6	1 5 5 5 8
7	2 3 4 5 8
8	0 1 1 5
9	0 2 3

Key: 5 | 2 means 52

Q6 I've been measuring the length of beetles for a science project. Here are the lengths in millimetres:

12 18 20 11 31
19 27 34 19 22

Complete the stem and leaf diagram on the right to show the results.

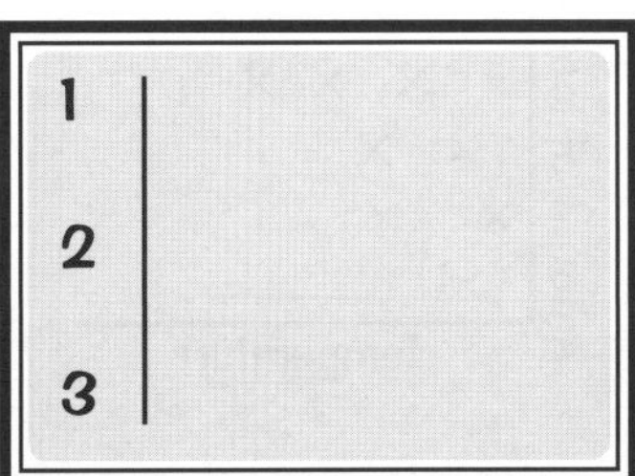

Key: 2 | 2 means 22

This is nothing new... you've seen it all before. All you do is read off the information from the chart.

Tables, Charts and Graphs

Q7 These are the shoe sizes and heights for 12 pupils in Year 11.

Shoe size	5	6	4	6	7	7	8	3	5	9	10	10
Height (cm)	155	157	150	159	158	162	162	149	152	165	167	172

On the grid below draw a scattergraph to show this information.

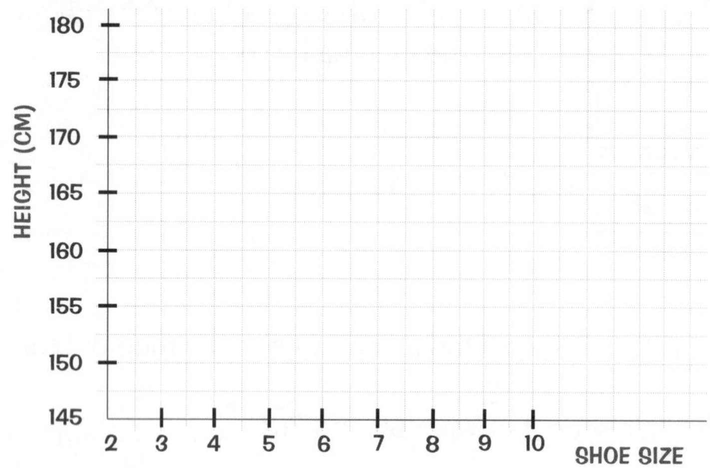

Draw a line of best fit on your scattergraph.

A line of best fit goes through the middle of the points.

What does the scattergraph tell you about the relationship between shoe size and height for these pupils?

....

Q8 The scattergraphs below show the relationship between:

a) The temperature of the day and the amount of ice cream sold.
b) The price of ice cream and the amount sold.
c) The age of customers and the amount of ice cream sold.

a) Amount of ice-cream / Temperature

b) Amount of ice-cream / Price

c) Amount of ice-cream / Age

Describe the correlation of each graph and say what each graph tells you.

a) ....

b) ....

c) ....

Tables, Charts and Graphs

Q9 This scattergraph shows how much time a group of teenagers spend on outdoor activities and playing computer games.

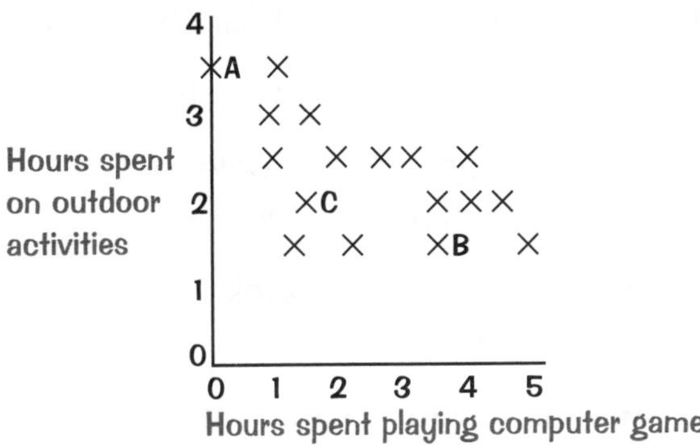

Which of the points A, B or C represent each of these statements?

The rugby practice was a long one so I didn't have much time to play on the computer.
I don't have a computer!
I went to visit a friend. We played a bit of football then spent most of the evening playing his new computer game.

a) Point ....

b) Point ....

c) Point ....

Q10 Alice wanted to buy a second hand car. She wrote down the ages and prices of 15 similarly-sized cars on sale locally. Draw a scattergraph for Alice's information.

Age of car (years)	Price (£)
4	4995
2	7995
3	6595
1	7995
5	3495
8	4595
9	1995
1	7695
2	7795
6	3995
5	3995
1	9195
3	5995
4	4195
9	2195

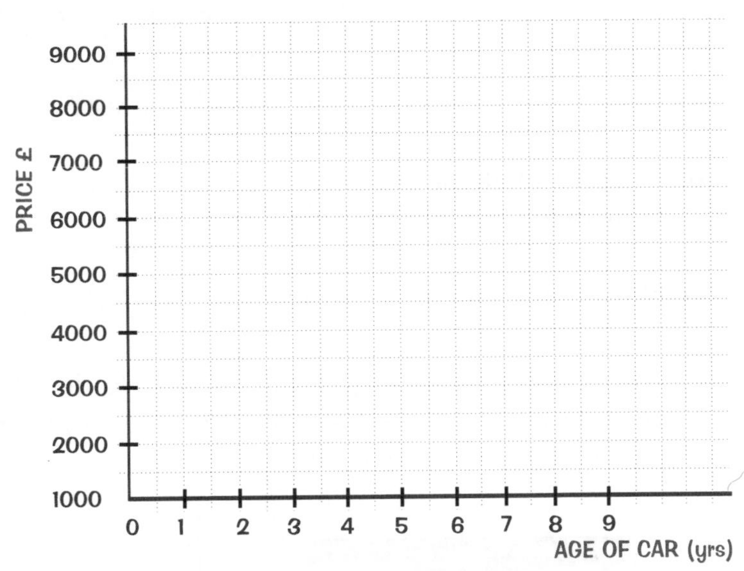

What does the scattergraph tell you about the relationship between the age of a car and its price?

....

The big word you're supposed to use in these questions is CORRELATION — and they're very keen on it, so make sure you know what it means.

Bar Charts

Q1 Here is a horizontal bar chart showing the favourite colours of a class of pupils.

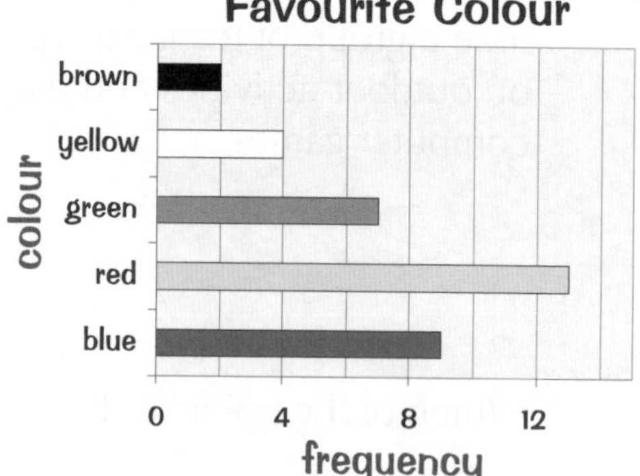

a) How many like blue best? ....

b) How many more people chose red than yellow? ....

c) How many pupils took part in this survey? ....

d) What fraction of the class prefer green? ....

Q2 This bar chart shows the marks from a test done by some students:

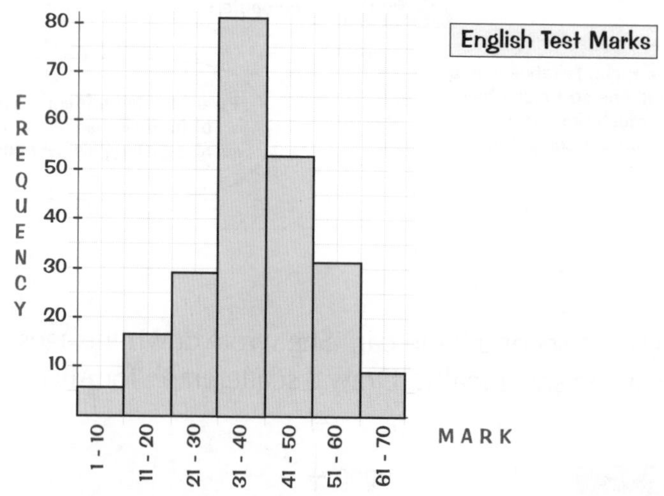

a) How many students scored 20 marks or less? ....

b) The pass mark for this test was 31.
How many students passed the test? ....

c) How many students took the test? ....

Q3 The local library is carrying out a survey to find the amount of time people spend reading each day, to the nearest minute. Complete the frequency table below and then draw a bar chart to show the results.

Time spent reading (mins)	Tally	Frequency
0 - 14	~~IIII~~ I	6
15 - 29	~~IIII~~ III	
30 - 44	III	
45 - 59	~~IIII~~	
60 - 74	III	

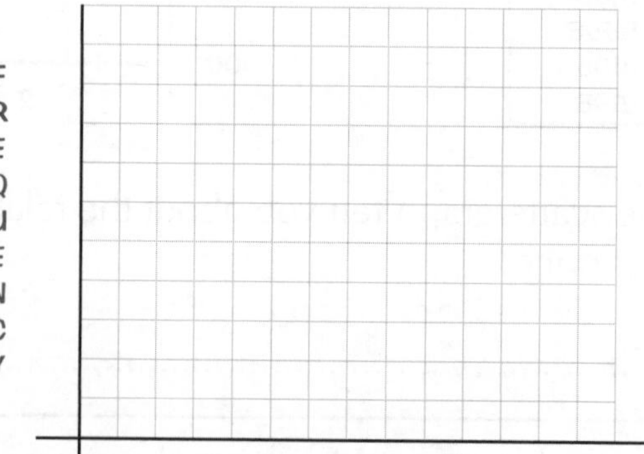

Pie Charts

Q1 This table shows the daily amount of air time of programme types on TV:

Programme	Hours	Angle
News	5	75°
Sport	3	
Music	2	
Current Affairs	3	
Comedy	2	
Other	9	
Total	24	

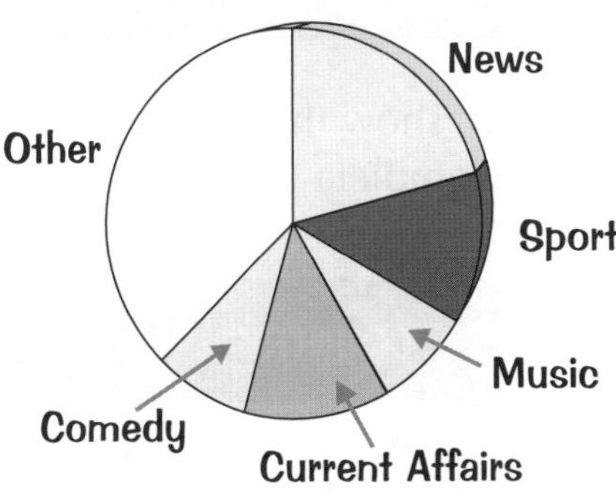

Using an angle measurer, complete the table by finding the size of the angle represented by each type of programme. The first angle is done for you.

Q2 Sandra is giving a presentation on her company's budget. She has decided to present the budget as a pie chart. The company spends £54 000 each week on various items, which are shown on the pie chart below.

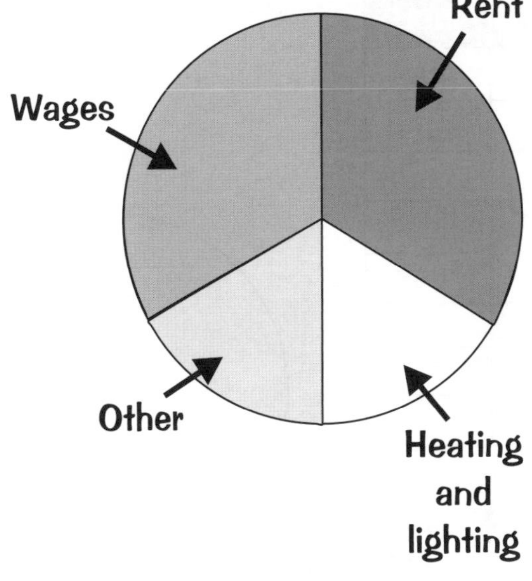

a) What fraction of the budget is spent on wages each week?

....

b) How much money is spent on wages each week?

....

c) What fraction of the budget is spent on heating and lighting each week?

....

d) How much money is spent on heating and lighting each week?

....

Pie Charts

Q3 In a University department there are 180 students from different countries.

Country	UK	Malaysia	Spain	Others
Number of students	90	35	10	45

To show this on a pie chart you have to work out the angle of each sector. Complete the table showing your working. The UK is done for you.

COUNTRY	WORKING	ANGLE in degrees
UK	$90 \div 180 \times 360 =$	180°
MALAYSIA		
SPAIN		
OTHERS		

Now complete the pie chart using an angle measurer. The UK sector is done for you.

Q4 Pupils at a school were asked about their activities at weekends. The results are shown in the table. Complete the table and then draw the pie chart using an angle measurer.

ACTIVITY	HOURS	WORKING	ANGLE
Homework	6	$6 \div 48 \times 360 =$	45°
Sport	2		
TV	10		
Computer games	2		
Sleeping	18		
Listening to music	2		
Paid work	8		
Total	48		

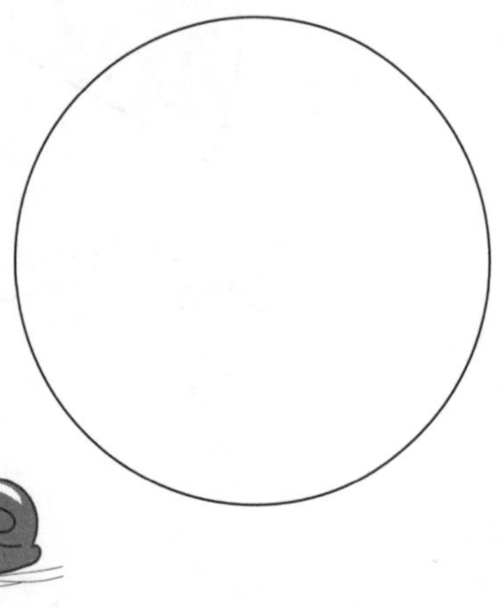

The full circle (that's all 360° of it) represents the total of everything — so you shouldn't find any gaps in it, basically.

Probability

When you're asked to find the probability of something, always check that your answer is between 0 and 1. If it's not, you know straight away that you've made a mistake.

Q1 Write down whether these events are impossible, unlikely, evens, likely or certain.

a) It will rain somewhere in the UK this month. ....

b) You will live to be 300 years old. ....

c) The next baby to be born in Hemel Hempstead will be female. ....

d) Next year April will be the month after March. ....

Q2 A bag contains ten balls. Five are red, three are yellow and two are green. What is the probability of picking out:

a) A yellow ball? ....

b) A red ball? ....

c) A green ball? ....

d) A red or a green ball? ....

e) A blue ball? ....

Q3 Write down the probability of these events happening:

a) Throwing an odd number on a fair six-sided dice. ....

b) Drawing a black card from a standard pack of playing cards. ....

c) Drawing a black King from a standard pack of cards. ....

d) Throwing a prime number on a fair six-sided dice. ....

Q4 Mike and Nick play a game of pool. The probability of Nick winning is 7/10.

a) Put an arrow on the probability line below to show the probability of Nick winning. Label this arrow N.

b) Now put an arrow on the probability line to show the probability of Mike winning the game. Label this arrow M.

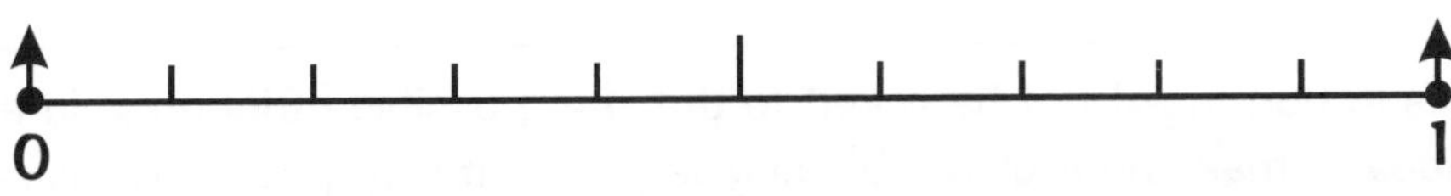

Probability

Q5 The outcome when a coin is tossed is head (H) or tail (T).
Complete this table of outcomes when two coins are tossed one after the other.

a) How many possible outcomes are there? ....

b) What is the probability of getting 2 heads? ....

c) What is the probability of getting a head followed by a tail? ....

		2nd COIN	
		H	T
1st COIN	H		
	T		

Q6 Lucy has designed a game for the school fair. Two dice are rolled at the same time and the scores on the dice are added together. She needs to work out the probability of all the possible outcomes so she can decide which outcomes will get a prize.
Complete the table of possible outcomes below.

How many different combinations are there? ....

		SECOND DICE					
		1	2	3	4	5	6
FIRST DICE	1						
	2	3					
	3						
	4						
	5			8			
	6						

What is the probability of scoring:

a) 2 ....

b) 6 ....

c) 10 ....

d) More than 9 ....

e) Less than 4 ....

f) More than 12 ....

g) Lucy would like to give a prize for any outcome that has a 50% chance of occurring.
Suggest an outcome that has a 50% chance of occurring.

....

Q7 Two spinners are spun and the scores are multiplied together.

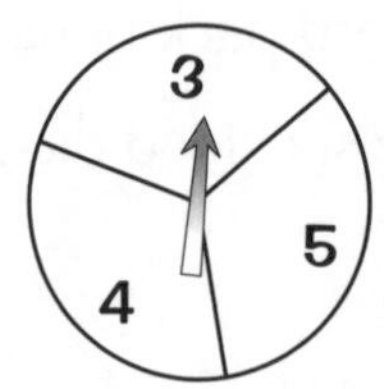

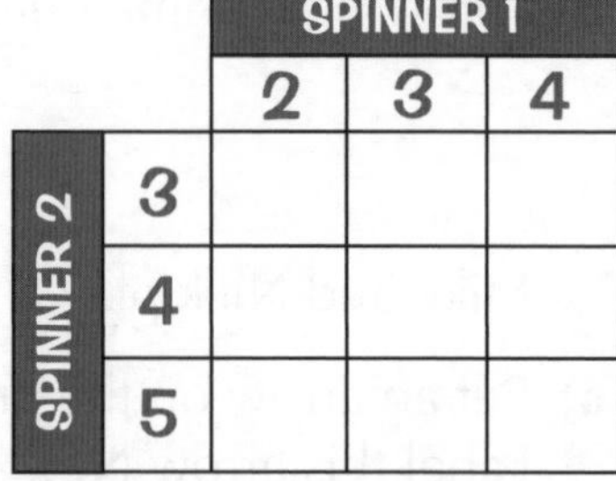

		SPINNER 1		
		2	3	4
SPINNER 2	3			
	4			
	5			

Fill in this table of possible outcomes.

What is the probability of scoring 12? ....

To win you have to score 15 or more.
What is the probability of winning? ....

In the exam, you might not be asked to put the "possible outcomes" in a table. But it's a good idea to make your own table anyway — that way you don't miss any out.

Probability

Well, OK, the probability is that you'd rather not be doing these at all... still — this is the last page, so I'm sure you'll cope for a bit longer.

Q8 One day Sarah did a survey in her class on sock colour. She found out that pupils were wearing white socks, black socks or red socks. Jack said "If I pick someone at random from the class, then the probability that they are wearing red socks is 1/3." Explain why Jack might be wrong.

....

....

Q9 Imagine you have just made a 6-sided spinner in Design and Technology. How could you test whether or not it's a fair spinner?

Remember — if the spinner's fair, the probability of landing on each side is the same.

....

....

....

Q10 a) A biased dice was rolled 40 times. A six came up 14 times. Calculate the relative frequency that a six was rolled.

....

b) The same dice was rolled another 60 times. From this, a six came up 24 times. Calculate the relative frequency that a six was rolled.

....

c) Use the data from **a)** and **b)** to make the best estimate you can of the probability of rolling a six with the dice.

....

Q11 "There is a 50% chance that it will rain tomorrow because it will either rain or it won't rain." Is this statement true or false? Explain your answer.

....

....

....

Ordering Numbers and Place Value

You may as well put your calculator away right now — you're not allowed to use one anywhere in Unit 2.

no calculators!!

Q1 Put these numbers in ascending (smallest to biggest) order.

a) 23 117 5 374 13 89 67 54 716 18

....

b) 1272 231 817 376 233 46 2319 494 73 1101

....

Q2 Lewis is writing a cheque for £278.04.
Write this amount in words as it should be written on the cheque.

....

Q3 Write down the value of the number 4 in each of these.

For example: 408 *hundreds*

a) 347 .... **b)** 41 .... **c)** 5478 ....

d) 6754 .... **e)** 4897 .... **f)** 6045 ....

g) 64 098 .... **h)** 745 320 .... **i)** 405 759 ....

j) 2 402 876 .... **k)** 4 987 321 .... **l)** 6 503 428 ....

Q4 Put these values in order of size — from the smallest to the largest.

a) 3.42 4.23 2.43 3.24 2.34 4.32

....

b) 6.7 6.704 6.64 6.642 6.741

....

Always look at the whole number part first, then the first digit after the decimal point, then the next etc.

c) 1002.8 102.8 1008.2 1020.8 108.2

....

d) £400.20 £402.22 £402.02 £400.02 £402.20

....

Negative Numbers

Q1 Write these numbers in the correct position on the number line below:

a) –4 3 2 –3 –5 1

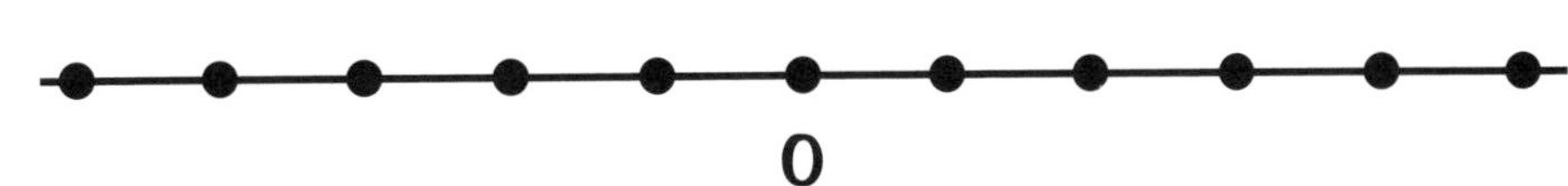

b) Which temperature is lower (colder), 8 °C or –4 °C ? ....

c) Which temperature is 1° warmer than –24 °C ? ....

Put the correct symbol, < or >, between the following pairs of numbers:

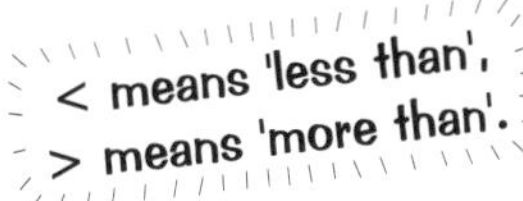

d) 4 –8

e) –6 –2

f) –8 –7

g) –3 –6

h) –1 1

i) –3.6 –3.7

j) Rearrange the following numbers in order of size, largest first:

–2 2 0.5 –1.5 –8 ....

k) The weather forecast says that the current temperature is –3 °C, but could reach –5 °C in the next few hours. Does this mean that it is going to get colder or warmer?

....

l) The next day, the forecast says that the temperature is currently –4 °C and will fall by 5 °C overnight. What is the forecast overnight temperature?

....

Always draw a number line to count along, so you can see what you're doing.

Multiplying by 10, 100, etc.

Multiplying by 10, 100 or 1000 moves each digit 1, 2 or 3 places to the left — you just fill the rest of the space with zeros.

Fill in the missing numbers. Remember, do not use a calculator for this page.

Q1 **a)** 6 × ☐ = 60 **b)** 0.07 × ☐ = 0.7

c) 6 × ☐ = 600 **d)** 0.07 × ☐ = 7

e) 6 × ☐ = 6000 **f)** 0.07 × ☐ = 70

Q2 Which number is ten times as large as 25? ....

Q3 Which number is one hundred times as large as 93? ....

Q4 **a)** 8 × 10 = ... **b)** 34 × 100 = ... **c)** 52 × 100 = ...

d) 9 × 1000 = ... **e)** 436 × 1000 = ... **f)** 0.2 × 10 = ...

g) 6.9 × 10 = ... **h)** 4.73 × 100 = ...

Q5 For a school concert chairs are put out in rows of 10.
How many will be needed for 16 rows? ....

Q6

How much do 10 chickens cost? ....

Q7 A school buys calculators for £2.45 each.
How much will 100 cost? ....

Q8 **a)** 20 × 30 = ... **b)** 40 × 700 = ... **c)** 250 × 20 = ...

d) 6000 × 210 = ... **e)** 18 000 × 500 = ...

Q9

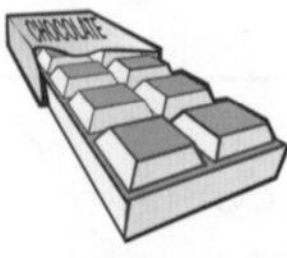

A shop bought 700 bars of chocolate for £0.43 each.
How much did they cost altogether?

....

Dividing by 10, 100, etc.

Dividing by 10, 100 or 1000 moves each digit 1, 2 or 3 places to the right.

Answer these questions <u>without</u> using a calculator:

Q1 **a)** 30 ÷ 10 = ... **b)** 43 ÷ 10 = ... **c)** 5.8 ÷ 10 = ...

d) 63.2 ÷ 10 = ... **e)** 0.5 ÷ 10 = ... **f)** 400 ÷ 100 = ...

g) 423 ÷ 100 = ... **h)** 228.6 ÷ 100 = ... **i)** 61.5 ÷ 100 = ...

j) 2.96 ÷ 100 = ... **k)** 6000 ÷ 1000 = ... **l)** 6334 ÷ 1000 = ...

m) 753.6 ÷ 1000 = ... **n)** 8.15 ÷ 1000 = ... **o)** 80 ÷ 20 = ...

p) 860 ÷ 20 = ... **q)** 2400 ÷ 300 = ... **r)** 480 ÷ 40 = ...

s) 860 ÷ 200 = ... **t)** 63.9 ÷ 30 = ...

Q2 Ten people share a Lottery win of £62.
How much should each person receive?

....

Q3 Blackpool Tower is 158 m tall. A model of it is being built to a scale of 1 : 100.
How tall should the model be?

....

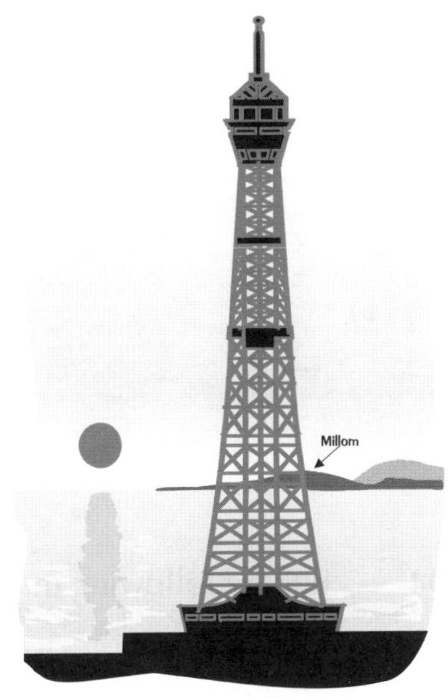

Q4 A box of 40 chocolates contains 56 g of saturated fat.
How much saturated fat is there in 1 chocolate?

....

Q5 Mark went on holiday to France. He exchanged £300 for 342 euros to spend while he was there. How many euros did he get for each £1?

....

Adding and Subtracting Decimals

Q1 Work out the answers without using a calculator.

a) 2.4 +3.2

b) 3.5 +4.6

c) 6.2 +5.9

d) 7.34 +6.07

e) 9.08 +4.93

f) 15.73 +25.08

g) 26.05 +72.95

Q2 Write these out in columns and work out the answers without using a calculator.

a) 3.6 + 7.3 **b)** 21.4 + 13.8 **c)** 0.9 + 5.6 **d)** 9.98 + 6.03 **e)** 2.9 + 7

f) 4.36 + 7.1 **g)** 9.8 + 1.05 **h)** 6 + 6.75 **i)** 0.28 + 18.5 **j)** 47.23 + 6.7

Q3 Work these out without a calculator:

a) 9.8 −3.1

b) 7.3 −2.3

c) 6.2 −1.5

d) 8.6 −3.9

e) 7.0 −1.6

f) 13.6 −12.7

g) 14.65 −4.7

h) 8.34 −4.65

Q4 Put the following in columns first, then work them out without using a calculator:

a) 8.5 – 1.6 **b)** 18.3 – 5.9 **c)** 24.1 – 16.3

d) 9 – 3.6 **e)** 40 – 2.3 **f)** 51 – 18.32

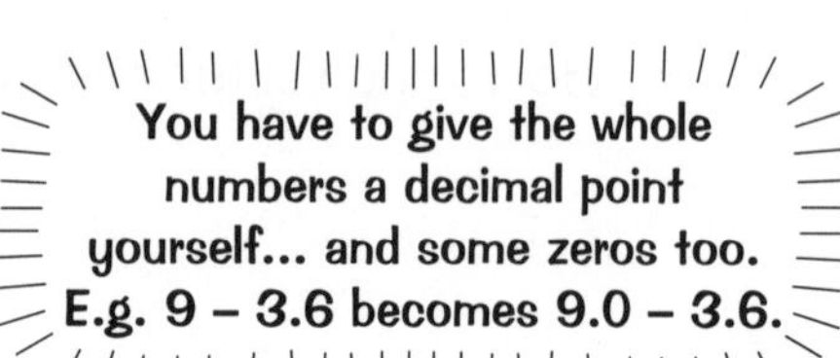

You have to give the whole
numbers a decimal point
yourself... and some zeros too.
E.g. 9 – 3.6 becomes 9.0 – 3.6.

Multiplying

Q1 Multiply the following without a calculator:

a) 23 × 2 = ...
b) 40 × 3 = ...
c) 53 × 4 = ...

d) 13 × 5 = ...
e) 25 × 4 = ...
f) 42 × 3 = ...

g) 18 × 2 = ...
h) 54 × 3 = ...
i) 75 × 5 = ...
j) 93 × 4 = ...

k) 308 × 4 = ...
l) 825 × 3 = ...
m) 346 × 5 = ...
n) 286 × 6 = ...

o) 126 × 14 = ...
p) 413 × 26 = ...
q) 309 × 61 = ...
r) 847 × 53 = ...

Q2 Without using a calculator, work out the total cost of:

a) 6 pens at 54p each.

....

b) 12 kg of potatoes at 76p per kg.

....

Q3 Without using a calculator, calculate how many hours there are in a year (365 days).

....

Dividing

Do these division questions without a calculator:

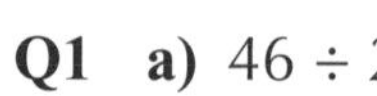

Q1 **a)** 46 ÷ 2 You may wish to set the sum out like this:

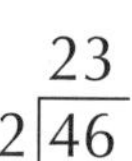

23
2|46

If the number doesn't divide exactly into the first digit, you have to carry the remainder over.

b) 86 ÷ 2 **c)** 96 ÷ 3 **d)** 76 ÷ 4

e) 85 ÷ 5 **f)** 96 ÷ 6 **g)** 91 ÷ 7

Q2 Try these:

a) 834 ÷ 3 **b)** 645 ÷ 5 **c)** 702 ÷ 6

d) 900 ÷ 4 **e)** 780 ÷ 12 **f)** 378 ÷ 18

Q3 Seven people share the cost of a meal equally between them. The meal costs £112. How much does each person have to pay?

....

Q4 The table shows nutritional information for a cake. The cake is cut into 8 equal slices. How many calories (kcal) are in each slice?

....

	Per cake
Energy	944 kcal / 3952 kJ
Protein	13.6g
Fat	24.8g
(saturates)	(13.2g)
Carbohydrate	166g
(sugars)	(98g)

Multiplying Decimals

A good way to multiply decimals is to ignore the decimal point to start with — just multiply the numbers. Then put the point back in and CHECK your answer looks sensible.

Do these calculations without using a calculator:

Q1 **a)** 3.2×4 **b)** 8.3×5 **c)** 6.4×3

= .... = ... = ...

d) 21×0.3 **e)** 263×0.2 **f)** 2.4×3.1

= .... = ... = ...

Q2 At a petrol station each pump shows a price table as shown below. Complete the table when the cost of unleaded petrol is 94.9p per litre.

Litres	Cost in pence
1	94.9
5	
10	
20	
50	

Q3 Jason can run 5.3 metres in 2 seconds. How far will he run if he keeps up this pace for:

a) 20 seconds **b)** 60 seconds **c)** 5 minutes

....

Q4 Steve is ordering some new garden fencing online. He knows what size he wants in yards, feet and inches, but the website gives sizes in centimetres. Using the information in the table, work out:

a) The number of centimetres in 3 inches.

....

b) The number of centimetres in 6 feet.

....

c) The number of centimetres in 45 yards.

....

Conversions

1 inch = 2.54 centimetres

1 foot = 12 inches

1 yard = 3 feet

1 metre = 3.28 feet

Dividing Decimals

This is the same thing, really — you divide the numbers first, then you put in the point where it should be.

Q1 Divide these without a calculator.

a) 8.4 ÷ 2 You may wish to set the sum out like this: $2\overline{)8.4}$ with 4.2 written above

b) 7.5 ÷ 3 **c)** 8.5 ÷ 5 **d)** 26.6 ÷ 7

e) 6.2 ÷ 5 **f)** 2.3 ÷ 4 **g)** 0.9 ÷ 5

Q2 Try these without a calculator:

a) 2.06 ÷ 8 **b)** 0.405 ÷ 6 **c)** 0.3 ÷ 8

d) 4.75 ÷ 5 **e)** 8.28 ÷ 9 **f)** 0.944 ÷ 8

No calculators allowed for these two either:

Q3 An 8 kg bag of rice costs £9.36. How much does the rice cost per kg?

....

Q4 A joiner drills six equally spaced holes in a length of wood. The first and last holes are 6.95 m apart. What is the distance between each hole?

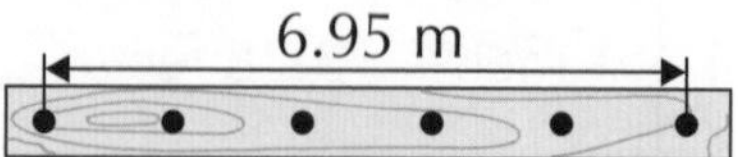

....

Prime Numbers

Basically, prime numbers don't divide by anything (except 1 and themselves).
5 is prime — it will only divide by 1 or 5. 1 is an exception to this rule — it is not prime.

Q1 Write down the first ten prime numbers. ....

Q2 Give a reason for 27 not being a prime number. ....

Q3 Using any or all of the figures **1, 2, 3, 7** write down:

a) the smallest prime number ....

b) a prime number greater than 20 ....

c) a prime number between 10 and 20 ....

d) two prime numbers whose sum is 20 .... , ...

e) a number that is not prime ....

Q4 Find all the prime numbers between 40 and 50. ....

Q5 In the ten by ten square opposite, ring all the prime numbers.

The first three have been done for you.

1	2	3	4	5	6	7	8	9	10
11	12	13	14	15	16	17	18	19	20
21	22	23	24	25	26	27	28	29	30
31	32	33	34	35	36	37	38	39	40
41	42	43	44	45	46	47	48	49	50
51	52	53	54	55	56	57	58	59	60
61	62	63	64	65	66	67	68	69	70
71	72	73	74	75	76	77	78	79	80
81	82	83	84	85	86	87	88	89	90
91	92	93	94	95	96	97	98	99	100

Q6 A school ran three evening classes: judo, karate and kendo.
The judo class had 29 pupils, the karate class had 27 and the kendo class 23.
For which classes would the teacher have difficulty dividing the pupils into equal groups?

....

Q7 Find three sets of three prime numbers which add up to the following numbers:

10 ...., ..., ...   **29** ...., ..., ...   **41** ...., ..., ...

Multiples

The multiples of a number are its times table — if you need multiples of more than one number, do them separately then pick the ones in both lists.

Q1 What are the first five multiples of:

a) 4? ....

b) 7? ....

c) 12? ....

d) 18? ....

Q2 Find a number which is a multiple of:

A quick way to do these is just to multiply the numbers together.

a) 2 and 6 ....

b) 7 and 5 ....

c) 2 and 3 and 7 ....

d) 4 and 5 and 9 ....

Q3 Steven is making cheese straws for a dinner party. There will be **either** six or eight people at the party. He wants to be able to share the cheese straws out equally. How many cheese straws should he make?

There's more than one right answer to this question.

....

Q4 a) Find a number which is a multiple of 3 and 8 ....

b) Find another number which is a multiple of 3 and 8 ....

c) Find another number which is a multiple of 3 and 8 ....

Q5 Which of these numbers 14, 20, 22, 35, 50, 55, 70, 77, 99 are multiples of:

a) 2? .... **b)** 5? ....

c) 7? .... **d)** 11? ....

Factors

Factors multiply together to make other numbers.
E.g. <u>1 × 6 = 6</u> and <u>2 × 3 = 6</u>, so <u>6 has factors 1, 2, 3 and 6</u>.

Q1 List the factors of the following numbers.
Each factor should be written once, with no repeats.

a) 18 ....
b) 22 ....
c) 35 ....
d) 7 ....
e) 16 ....
f) 49 ....
g) 32 ....
h) 31 ....
i) 50 ....
j) 62 ....
k) 81 ....
l) 100 ....

Q2 **a)** I am a factor of 24.
I am an odd number.
I am greater than 1.
What number am I?
....

b) I am a factor of 30.
I am an even number.
I am less than 5.
What number am I?
....

Q3 Circle all the factors of 360 in this list of numbers.

      7 8  

Q4 48 students went on a geography field trip. Their teachers split them into equal groups.
Suggest five different ways that the teachers might have split up the students:

.... groups of ....
.... groups of ....
.... groups of ....
.... groups of ....
.... groups of ....

Factors

Q5 …niel is tiling his bathroom using small square tiles. He wants to put a solid …ctangle of blue tiles over his bath. Daniel has 42 blue tiles in total.

a) How could Daniel arrange all 42 tiles to make the longest rectangle?

....

b) How could Daniel arrange all 42 tiles to make the squarest rectangle?

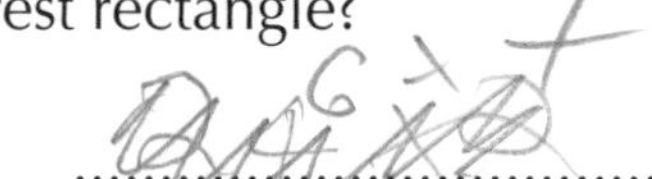

....

Q6 A perfect number is one where the factors add up to the number itself. For example, the factors of 28 are 1, 2, 4, 7 and 14 (not including 28 itself). These add up to 1 + 2 + 4 + 7 + 14 = 28, and so 28 is a perfect number.

Complete this table, and circle any perfect numbers in the left hand column.

Number	Factors (excluding the number itself)	Sum of Factors
2		
4	1, 2	3
6		
8		
10		

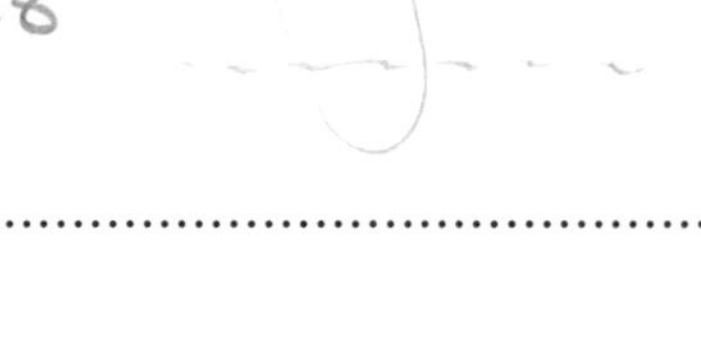

The sum of the factors is all the factors added together.

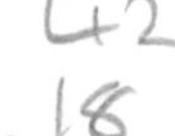

Q7 a) What is the biggest number that is a factor of both 42 and 18?

....

b) What is the smallest number that has both 4 and 18 as factors?

....

Q8 Complete the factor trees below to express each number as a product of prime factors. The first one has been done for you.

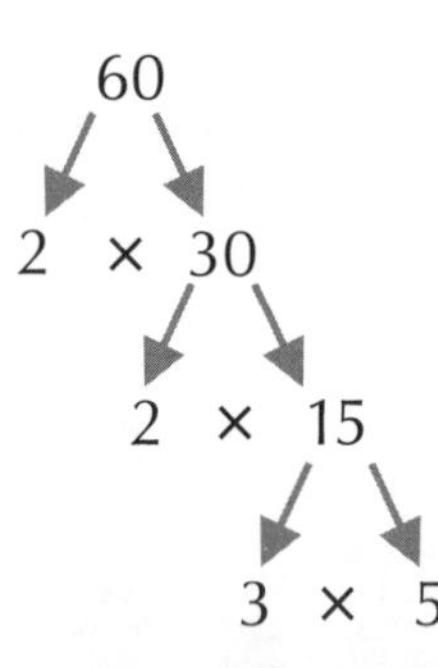

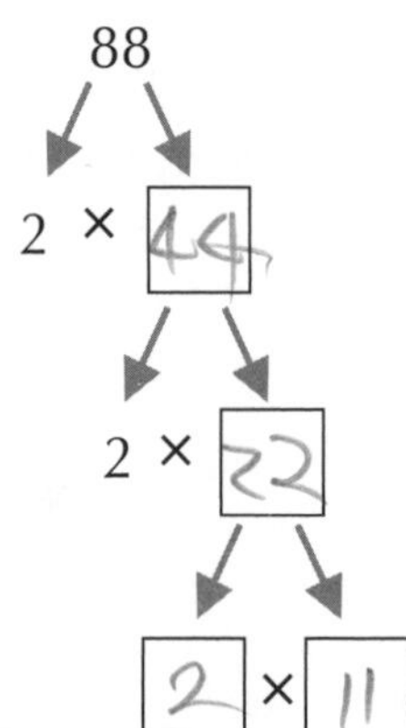

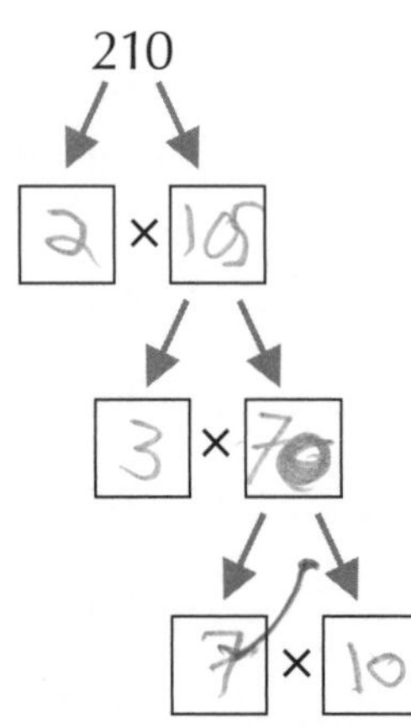

60 = 2×2×3×5

88 = 2×2×..........×..........

210 = ...×..........×..........×..........

LCM and HCF

The Lowest Common Multiple (LCM) is the **SMALLEST** number that will **DIVIDE BY ALL** the numbers in question.

The Highest Common Factor (HCF) is the **BIGGEST** number that will **DIVIDE INTO ALL** the numbers in question.

Q1 **a)** List the first ten multiples of 6, starting at 6.

....

b) List the first ten multiples of 5, starting at 5.

....

c) What is the LCM of 5 and 6? ....

Q2 For each set of numbers find the HCF.

a) 3, 5 ....

b) 6, 8 ....

c) 10, 15 ....

d) 27, 48 ....

e) 14, 21 ....

f) 11, 33, 121 ....

Q3 For each set of numbers, find the LCM.

a) 3, 5 ....

b) 6, 8 ....

c) 10, 15 ....

d) 15, 18 ....

e) 14, 21 ....

f) 11, 33, 44 ....

Q4 Lars, Rita and Alan regularly go swimming. Lars goes every 2 days, Rita goes every 3 days and Alan goes every 5 days. They all went swimming together on Friday 1st June.

This is an LCM question in disguise.

a) On what date will Lars and Rita next go swimming together?

....

b) On what date will Rita and Alan next go swimming together?

....

c) On what day of the week will all 3 next go swimming together?

....

d) Which of the 3 (if any) will go swimming on 15th June?

....

Powers

Q1 Complete the following:

a) $2^4 = 2 \times 2 \times 2 \times 2 =$ ....

b) $10^3 = 10 \times 10 \times 10 =$ ....

c) $3^5 = 3\times$ .... = ...

d) $4^6 = 4 \times$ .... = ...

e) $1^9 = 1 \times$ .... = ...

f) $5^6 = 5 \times$ .... = ...

Q2 Simplify the following:

a) $2 \times 2 \times 2 \times 2 \times 2 \times 2 \times 2 \times 2 =$ ....

b) $12 \times 12 \times 12 \times 12 \times 12 =$ ....

c) $m \times m \times m =$ ....

d) $y \times y \times y \times y =$ ....

Q3 Complete the following (the first one has been done for you):

a) $10^2 \times 10^3 = (10\times10) \times (10\times10\times10) = 10^5$

b) $10^3 \times 10^4 =$ .... =

c) $10^4 \times 10^2 =$ .... =

d) What is the quick method for writing down the final result in **b)** and **c)**?

....

Q4 Complete the following (the first one has been done for you):

a) $2^4 \div 2^2 = \dfrac{(2 \times 2 \times 2 \times 2)}{(2 \times 2)} = 2^2$

b) $2^5 \div 2^2 = \dfrac{(2 \times 2 \times 2 \times 2 \times 2)}{(2 \times 2)} =$ ....

c) $4^5 \div 4^3 = \dfrac{(4 \times 4 \times 4 \times 4 \times 4)}{\text{.......................}} =$ ....

d) $8^5 \div 8^2 = \dfrac{\text{.......................}}{\text{.......................}} =$ ....

e) What is the quick method for writing down the final result in **b)**, **c)** and **d)**?

....

You can always write out all the 10s or whatever if you get stuck.

Q5 Write the following as a single term:

a) $10^6 \div 10^4 =$ ....

b) $(8^2 \times 8^5) \div 8^3 =$ ....

c) $6^{10} \div (6^2 \times 6^3) =$ ....

d) $x^2 \times x^3 =$ ....

e) $a^5 \times a^4 =$ ....

f) $p^4 \times p^5 \times p^6 =$ ....

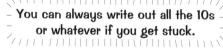

Powers are just a way of writing numbers in shorthand — they come in especially handy with big numbers. Imagine writing out 2^{138} — 2 × 2 ×... × 2 ×... × 2 × yawn × zzz...

Square and Cube Roots

Square root just means "**WHAT NUMBER TIMES ITSELF (i.e. 2×2) GIVES...**"
The square roots of 64 are 8 and –8 because 8×8=64 and -8×-8=64.
Cube root means "**WHAT NUMBER TIMES ITSELF TWICE (i.e. 2×2×2) GIVES ...**"
The cube root of 27 is 3 because 3×3×3=27.
Square roots always have a + and – answer, cube roots only have 1 answer.

Q1 Without using a calculator, write down both answers to each of the following:

a) $\sqrt{4}$ = ...
b) $\sqrt{16}$ = ...
c) $\sqrt{9}$ = ...
d) $\sqrt{49}$ = ...
e) $\sqrt{25}$ = ...
f) $\sqrt{100}$ = ...
g) $\sqrt{144}$ = ...
h) $\sqrt{64}$ = ...
i) $\sqrt{81}$ = ...

Q2 Without using a calculator, find the value of the following:

a) $\sqrt[3]{64}$ = ...
b) $\sqrt[3]{27}$ = ...
c) $\sqrt[3]{125}$ = ...
d) $\sqrt[3]{1000}$ = ...
e) $\sqrt[3]{8}$ = ...
f) $\sqrt[3]{8000}$ = ...

Q3 A square rug has an area of 25 m^2. What is the length of an edge?

....

Q4 A solid cube puzzle has a volume of 125 cm^3. Find the length of one of its edges.

....

Q5 Nida is buying a small storage box online. She sees a cube box with a volume of 1000 cm^3. What is the length of each box edge?

....

Q6 A farmer is buying fencing to surround a square field of area 3600 m^2. What length of fencing does he need to buy?

....

Fractions, Decimals, %

DON'T FORGET — no calculator allowed for any of these questions.

Q1 Change these fractions to decimals:

a) $\frac{1}{2}$ .... **b)** $\frac{3}{4}$ .... **c)** $\frac{7}{10}$ .... **d)** $\frac{19}{20}$ ....

e) $\frac{1}{100}$ .... **f)** $\frac{3}{8}$ .... **g)** $\frac{2}{1000}$ .... **h)** $\frac{1}{3}$ ....

Q2 Change these fractions to percentages:

a) $\frac{1}{4}$ .... **b)** $\frac{3}{10}$ .... **c)** $\frac{4}{5}$ .... **d)** $\frac{12}{25}$ ....

e) $\frac{8}{100}$ .... **f)** $\frac{2}{40}$ .... **g)** $\frac{7}{8}$ .... **h)** $\frac{11}{30}$ ....

Q3 Change these decimals to percentages:

a) 0.62 .... **b)** 0.74 .... **c)** 0.4 .... **d)** 0.9 ....

e) 0.07 .... **f)** 0.02 .... **g)** 0.125 .... **h)** 0.987 ....

Q4 Change these percentages to decimals:

a) 25% .... **b)** 49% .... **c)** 3% .... **d)** 30% ....

Q5 Change these percentages to fractions (in their lowest terms):

a) 75% .... **b)** 60% .... **c)** 15% .... **d)** 53% ....

Q6 Change these decimals to fractions (in their lowest terms):

a) 0.5 .... **b)** 0.8 .... **c)** 0.19 .... **d)** 0.25 ....

e) 0.64 .... **f)** 0.06 .... **g)** 0.125 .... **h)** 0.075 ....

A FRACTION IS A DECIMAL IS A PERCENTAGE — they're all just different ways of saying "a bit of" something.

Fraction Arithmetic

Fraction arithmetic becomes a nice cruise in the park, once you've learned the rules for dealing with each type of sum.

MULTIPLYING: just multiply the tops and the bottoms.
DIVIDING: turn the second fraction upside down, then multiply them.
ADDING OR SUBTRACTING:
1) Make the bottom number the same (get "a common denominator")
2) Add or subtract the top numbers only.

Answer the following questions without using a calculator.

Q1 Which is bigger?

a) 1/5 or 2/10 ....
b) 3/7 or 6/21 ....

c) 10/15 or 4/6 ....
d) 1/3 or 33/100 ....

For some of the questions below, you'll need to change the mixed fractions into top-heavy fractions at the start.

Q2 Do the following multiplications, expressing the answers as fractions in their lowest terms:

a) $\frac{4}{3} \times \frac{3}{4}$ ....
d) $2\frac{1}{2} \times \frac{3}{5}$ ....

b) $\frac{2}{5} \times \frac{3}{4}$ ....
e) $10\frac{2}{7} \times \frac{7}{9}$ ....

c) $\frac{11}{9} \times \frac{6}{5}$ ....
f) $2\frac{1}{6} \times 3\frac{1}{3}$ ....

Q3 Carry out the following divisions, and express each answer in its lowest terms:

a) $\frac{1}{4} \div \frac{3}{8}$ ....
d) $1\frac{1}{2} \div \frac{5}{12}$ ....

b) $\frac{1}{9} \div \frac{2}{3}$ ....
e) $10\frac{4}{5} \div \frac{9}{10}$ ....

c) $\frac{15}{24} \div \frac{6}{5}$ ....
f) $3\frac{7}{11} \div 1\frac{4}{11}$ ....

Q4 Add the two fractions, giving your answer as a fraction in its lowest terms:

a) $\frac{7}{8} + \frac{3}{8}$ ....
b) $\frac{1}{12} + \frac{3}{4}$ ....
c) $\frac{1}{3} + \frac{3}{4}$ ....

d) $1\frac{2}{5} + 2\frac{2}{3}$ ....
e) $\frac{1}{6} + 4\frac{1}{3}$ ....
f) $1\frac{3}{10} + \frac{2}{5}$ ....

Q5 Evaluate, giving your answer as a fraction in its lowest terms:

a) $\frac{11}{4} - \frac{2}{3}$ ....
b) $10 - \frac{2}{5}$ ....
c) $1\frac{3}{4} - 1\frac{1}{5}$ ....

d) $4\frac{2}{3} - \frac{7}{9}$ ....
e) $3\frac{1}{2} - \frac{2}{3}$ ....
f) $8 - \frac{1}{8}$ ....

Ratio

Q1 Divide the following quantities in the given ratio.

For example:

£400 in the ratio 1 : 4   1 + 4 = 5   £400 ÷ 5 = £80

1 × £80 = £80 and 4 × £80 = £320   £80 : £320

a) 100 g in the ratio 1 : 4   .... + ... = ... ÷ .... = ...

.... × .... = ... and .... × .... = ...

= .... : ....

b) 500 m in the ratio 2 : 3 = ... : ....

c) £12 000 in the ratio 1 : 2 = ... : ....

d) 6.3 kg in the ratio 3 : 4 = ... : ....

e) £8.10 in the ratio 4 : 5 = ... : ....

Q2 Now try these...

a) Adam and Mags win £24 000. They split the money in the ratio 1 : 5. How much does Adam get?

....

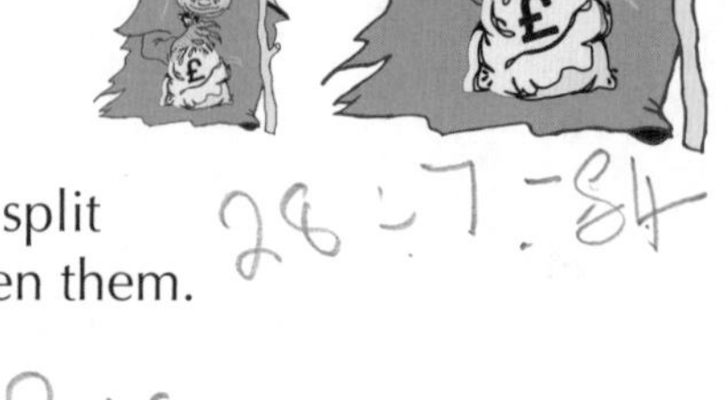

b) Sunil and Paul work in a restaurant. Any tips they earn are split in the ratio 3 : 4. One night they earned £28 in tips between them. Who got the most tip money? How much did they get?

.... got £......................

c) The total distance covered in a triathlon (swimming, cycling and running) is 15 km. It is split in the ratio 2 : 3 : 5. How far is each section?

Swimming = ..., cycling = ..., running = ...

A great way to check your answer works is to add up the individual quantities — they should add up to the original amount.

Algebra — Collecting Terms and Expanding

Algebra can be pretty scary at first. But don't panic — the secret is just to practise lots and lots of questions. Eventually you'll be able to do it without thinking, just like riding a bike. But a lot more fun, obviously...

<u>Simplifying</u> means collecting <u>like terms</u> together:	<u>Expanding</u> means removing brackets:
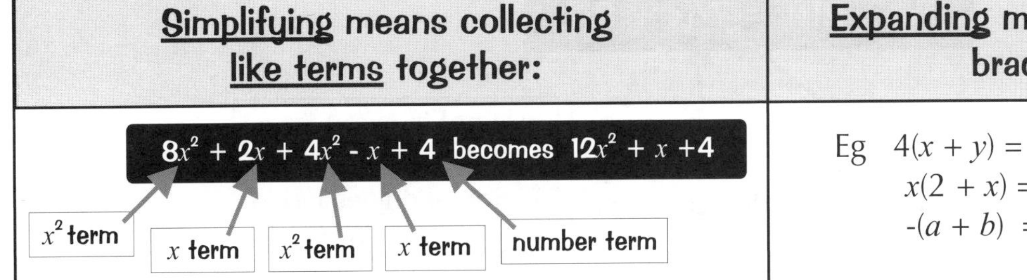	Eg $4(x + y) = 4x + 4y$ $x(2 + x) = 2x + x^2$ $-(a + b) = -a - b$

Q1 By collecting like terms, simplify the following. The first one is done for you.

a) $6x + 3x - 5 = 9x - 5$

b) $2x + 3x - 5x =$ ....

c) $9f + 16f + 15 - 30 =$ ....

d) $14x + 12x - 1 + 2x =$ ....

e) $3x + 4y + 12x - 5y =$ ....

f) $11a + 6b + 24a + 18b =$ ....

g) $9f + 16g - 15f - 30g =$ ....

h) $14a + 12a^2 - 3 + 2a =$ ....

Q2 Simplify the following. The first one is done for you.

a) $3x^2 + 5x - 2 + x^2 - 3x = 4x^2 + 2x - 2$

b) $5x^2 + 3 + 3x - 4 =$ ....

c) $13 + 2x^2 - 4x + x^2 + 5 =$ ....

d) $7y - 4 + 6y^2 + 2y - 1 =$ ....

e) $2a + 4a^2 - 6a - 3a^2 + 4 =$ ....

f) $15 - 3x - 2x^2 - 8 - 2x - x^2 =$ ....

g) $x^2 + 2x + x^2 + 3x + x^2 + 4x =$ ....

h) $2y^2 + 10y - 7 + 3y^2 - 12y + 1 =$ ....

Q3 Expand the brackets and then simplify if possible. The first one is done for you.

Careful with the minus signs — they multiply both terms in the bracket.

a) $2(x + y) = 2x + 2y$

b) $4(x - 3) =$ ....

c) $8(x^2 + 2) =$ ....

d) $-2(x + 5) =$ ....

e) $-(y - 2) =$ ....

f) $x(y + 2) =$ ....

g) $x(x + y + z) =$ ....

h) $8(a + b) + 2(a + 2b) =$ ....

Algebra — Taking out Common Factors

FACTORISING is just putting the brackets back in.
And when you've just spent all that time getting rid of them...

$$7x^2 + 21xy = 7x(x + 3y)$$

- largest number that will go into 7 and 21
- highest power of x, that will go into each term
- y is not in every term so it is not a common factor, and goes inside the brackets

Q1 Factorise the expressions below. Each has 4 as a common factor.

a) $4x + 8 =$ .... **c)** $4 - 16x =$ ....

b) $12 - 8x =$ .... **d)** $4x^2 + 64 =$ ....

Q2 Factorise the expressions below. Each has 7 as a common factor.

a) $21 - 7x =$ .... **c)** $14 + 21x =$ ....

b) $28x + 7 =$ .... **d)** $35x^2 - 14 =$ ....

Q3 Factorise the expressions below. Each has x as a common factor.

a) $2x + x^2 =$ .... **c)** $x - 16x^2 =$ ....

b) $2x - x^2 =$ .... **d)** $4x^2 - 3x =$ ....

Q4 Factorise the expressions below. Each has $2x$ as a common factor.

a) $2x + 4x^2 =$ .... **c)** $2x - 16x^2 =$ ....

b) $2x - 8x^3 =$ .... **d)** $4xy - 6x^2 =$ ....

Q5 Factorise the expressions below by taking out any common factors.

a) $2x + 4 =$ .... **f)** $30 + 10x =$ ....

b) $3x + 12 =$ .... **g)** $9x^2 + 3x =$ ....

c) $24 + 12x =$ .... **h)** $5x^2 + 10x =$ ....

d) $16x + 4y =$ .... **i)** $7x^2 + 21x =$ ....

e) $3x + 15 =$ .... **j)** $3y + xy^2 =$ ....

Formulas from Words

It's no big mystery — algebra is just like normal sums, but with the odd letter or two stuck in for good measure.

Q1 Write the algebraic expression for these:

a) Three more than x ....

b) Seven less than y ....

c) Four multiplied by x ....

d) y multiplied by y ....

e) Ten divided by b ....

f) A number add five ....

Q2 Steven is 16 years old. How old will he be in:

a) 5 years? .... **b)** 10 years? .... **c)** x years? ....

Q3 Tickets for a football match cost £25 each. What is the cost for:

a) 2 tickets? ....

b) 6 tickets? ....

c) y tickets? ....

CGP Wanderers Football Club
Vs United Rovers FC
Comfy Seat
East stand lower bit
Row 20
Seat 104
£25.00

Q4 There are n books in a pile. Write an expression for the number of books in a pile that has:

a) 3 more books ....

b) 4 fewer books ....

c) Twice as many books ....

Q5 a) This square has sides of length 3 cm.

What is its perimeter? ....

What is its area? ....

b) This square has sides of length d cm.

What is its perimeter? ....

What is its area? ....

3 cm
3 cm

See Unit 3 for more on area and perimeter.

d cm
d cm

Q6 The cost (C) of hiring a mountain bike is £10, plus £5 for each hour you use the bike (h). Write down a formula that can be used for working out the cost of hiring a bike.

....

Rearranging Formulas

Rearranging is getting the letter you want out of the formula and making it the subject.

Example:- Rearrange the formula $p = 3q + r$ to make q the subject.

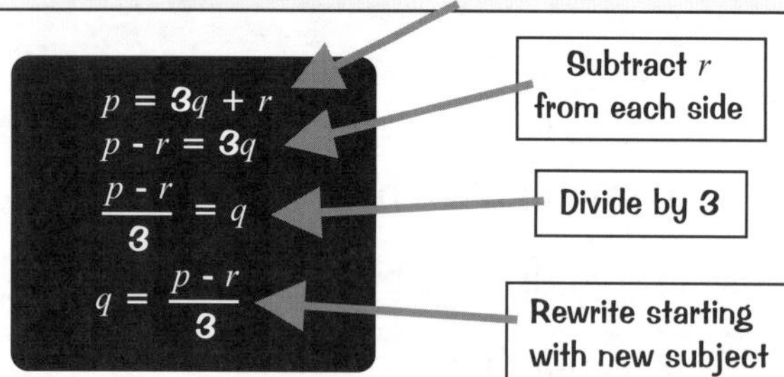

Remember
The same method applies to rearranging formulas as solving equations

Q1 Rearrange the following formulas to make the letter in brackets the new subject:

a) $y = x + 4$ (x) ....

b) $y = 2x + 3$ (x) ....

c) $y = 4x - 5$ (x) ....

d) $a = 7b + 10$ (b) ....

e) $w = 14 + 2z$ (z) ....

f) $s = 4t - 3$ (t) ....

g) $y = 3x + ½$ (x) ....

h) $y = 3 - x$ (x) ....

i) $y = 5(x + 2)$ (x) ....

Q2 Rearrange the following, to make the letter in brackets the subject of the formulas:

a) $y = \frac{x}{10}$ (x) ....

b) $s = \frac{t}{14}$ (t) ....

c) $a = \frac{2b}{3}$ (b) ....

d) $d = \frac{3e}{4}$ (e) ....

e) $f = \frac{3g}{8}$ (g) ....

f) $y = \frac{x}{5} + 1$ (x) ....

g) $y = \frac{x}{2} - 3$ (x) ....

h) $a = \frac{b}{3} - 5$ (b) ....

Q3 A car sales person is paid £w for working m months and selling c cars, where $w = 500m + 50c$.

a) Rearrange the formula to make c the subject. ....

b) Find the number of cars the sales person sells in 11 months if he earns £12,100 during that time. ....

Q4 The cost (c) of a taxi journey is £1.40 per mile (m), plus an extra charge of £1.50.

a) Write down a formula with c as the subject. ....

b) Rearrange this to make m the subject. ....

c) Frank wants to know how far his hotel is from the airport. He knows the taxi fare from one to the other is £9.90. How far is the hotel from the airport? ....

Substituting Values into Formulas

BODMAS — this funny little word helps you remember in which order to work formulas out. The example below shows you how to use it. Oh and by the way "Other" might not seem important, but it means things like powers and square roots, etc — so it is.

Example: if $z = \frac{x}{10} + (y - 3)^2$, find the value of z when $x = 40$ and $y = 13$.

1) Write down the formula with the numbers stuck in $z = \frac{40}{10} + (13 - 3)^2$,

2) Brackets first: $z = \frac{40}{10} + (10)^2$

3) Other next, so square: $z = \frac{40}{10} + 100$

4) Division before Addition: $z = 4 + 100$

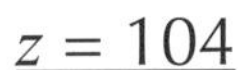

$\underline{z = 104}$

Q1 If $x = 3$ and $y = 6$ find the value of the following expressions.

a) $x + 2y$ .... **c)** $4(x + y)$ .... **e)** $2x^2$ ....

b) $2x \div y$ .... **d)** $(y - x)^2$ .... **f)** $2y^2$ ....

Q2 If $V = lwh$, find V when, **a)** $l = 7$, $w = 5$, $h = 2$ ....

b) $l = 12$, $w = 8$, $h = 5$ ....

Q3 Using the formula $z = (x - 10)^2$, find the value of z when,

a) $x = 20$ .... **b)** $x = 15$ .... **c)** $x = -1$ ....

Q4 If $V = u + at$, find the value of V when $u = 8$, $a = 9.8$ and $t = 2$. ....

Q5 The cost in pence (C) of hiring a sun lounger depends on the number (n) of hours you use it for, where $C = 100 + 30n$. Find C when,

a) $n = 2$ .... **b)** $n = 6$ .... **c)** $n = 3.5$ ....

Q6 The cost of framing a picture, C pence, depends on the dimensions of the picture. If $C = 10L + 5W$, where L is the length in cm and W is the width in cm, then find the cost of framing:

a) a picture of length 40 cm and width 24 cm ....

b) a square picture of sides 30 cm. ....

Q7 The time taken to cook a chicken is given as 20 minutes per lb plus 20 minutes extra. Find the time needed to cook a chicken weighing:

a) 4 lb ....

b) 7.5 lb ....

You need to write your own formula for this one.

Solving Equations

**You've got to get the letter on its own (x = ...).
You can add, divide... well, anything really — but you gotta do it to both sides or it'll all go horribly wrong.**

Q1 Solve these equations:

a) $a + 6 = 20$ ....

b) $b + 12 = 30$ ....

c) $48 + c = 77$ ....

d) $397 + d = 842$ ....

e) $e + 9.8 = 14.1$ ....

f) $3 + f = 7$ ....

Q2 Solve these equations:

a) $g - 7 = 4$ ....

b) $h - 14 = 11$ ....

c) $i - 38 = 46$ ....

d) $j - 647 = 353$ ....

e) $k - 6.4 = 2.9$ ....

f) $l - 7 = -4$ ....

Q3 Solve these equations:

a) $4m = 28$ ....

b) $7n = 84$ ....

c) $15p = 645$ ....

d) $279q = 1395$ ....

e) $6.4r = 9.6$ ....

f) $-5s = 35$ ....

Q4 Solve these equations:

a) $\frac{t}{3} = 5$ ....

b) $u \div 6 = 9$ ....

c) $\frac{v}{11} = 8$ ....

d) $\frac{w}{197} = 7$ ....

e) $x \div 1.8 = 7.2$ ....

f) $\frac{y}{-3} = 7$ ....

Q5 Melissa paid £23.40 for three identically-priced album downloads.

a) How much did each album cost? ....

b) Each album contained 12 songs. What was the price of each song? ....

Solving Equations

These are just like the last page... only there's an extra step. It's a good idea to get rid of any fractions before you do anything else.

Q6 Solve these equations:

a) $3x + 2 = 14$ ....

b) $5x - 4 = 31$ ....

c) $8 + 6x = 50$ ....

d) $20 - 3x = -61$ ....

Q7 Solve these equations:

a) $\frac{x}{3} + 4 = 10$ ....

b) $\frac{x}{5} - 9 = 6$ ....

c) $4 + \frac{x}{9} = 6$ ....

d) $\frac{x}{17} - 11 = 31$ ....

For help with these, look back over expanding brackets and collecting like terms.

Q8 Solve these equations:

a) $3(2x + 1) = 27$ ....

b) $2(4x + 1) + x = 56$ ....

c) $5x + 3 = 2x + 15$ ....

d) $2(x + 7) = 6x - 10$ ....

Q9 Solve the following:

a) $3(7 - 2x) = 2(5 - 4x)$ ....

b) $4(3x + 2) + 3 = 3(2x - 5) + 2$ ....

c) $6(x + 2) + 4(x - 3) = 50$ ....

d) $10(x + 3) - 4(x - 2) = 7(x + 5)$ ....

Q10 Florence booked 5 tickets for a concert. Each ticket cost £18.50 plus a booking fee. There was also a fee of £5 added to the order to cover postage.
The total amount Florence paid was £116.
Work out the booking fee for each ticket.

....

Inequalities

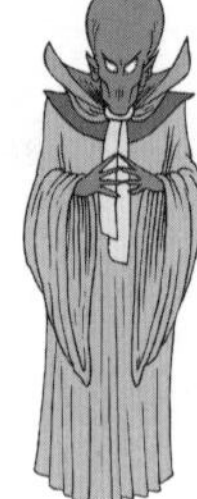

Yet another one of those bits of Maths that looks worse than it is — these are just like equations, really, except for the symbols.

The 4 Inequality Symbols:

$>$ means greater than	$<$ means less than
$\geqslant$ means greater than or equal to	$\leqslant$ means less than or equal to

Inequalities can be represented on number lines. You need to know this notation, too:

E.g.

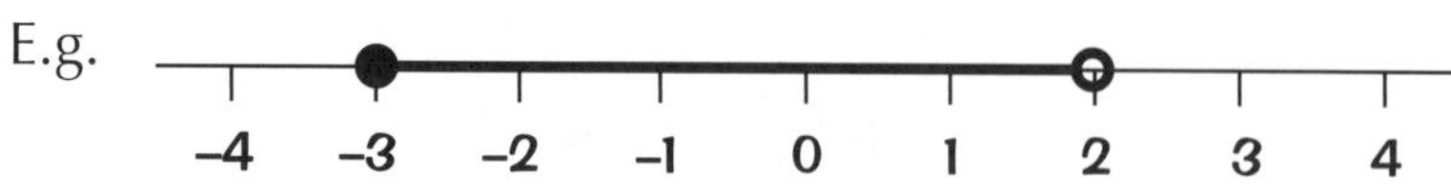

represents the inequality $-3 \leqslant x < 2$

REMEMBER:
- ● includes the value
- ○ does not include it

Q1 Write down an inequality for each of the diagrams below.

a)  ....

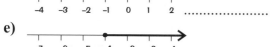

b) 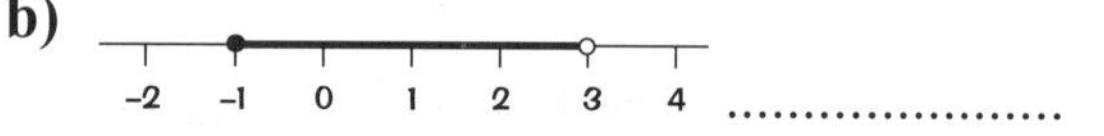 ....

c) 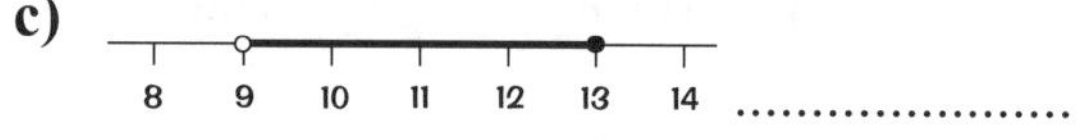 ....

d) 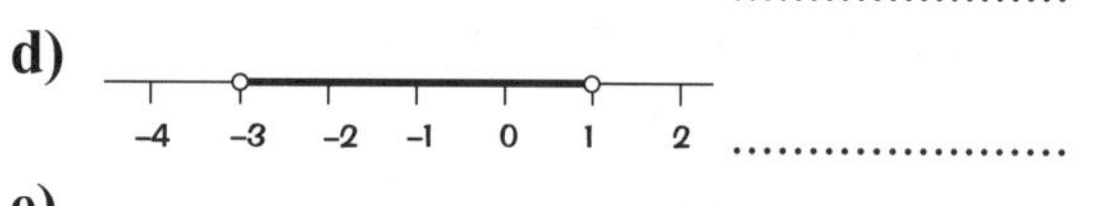 ....

e) ....

f) 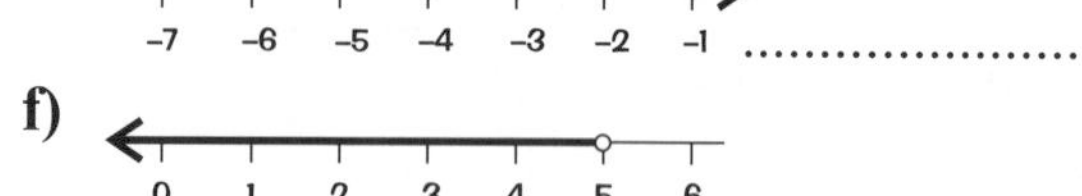 ....

g) 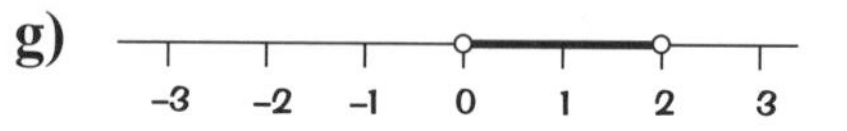 ....

h)  ....

i) 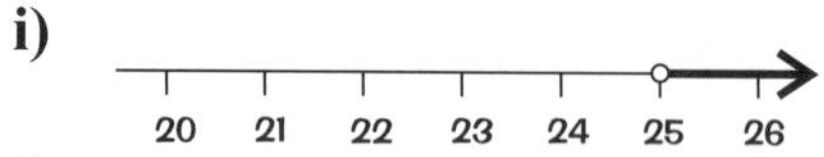 ....

j) 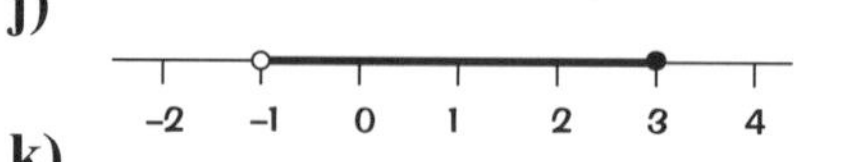 ....

k) ....

l) 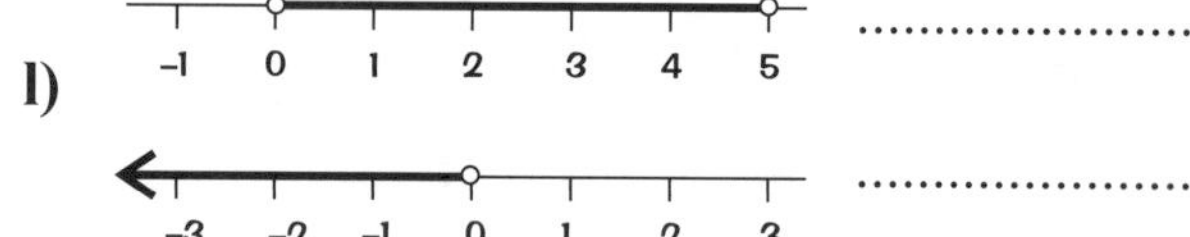 ....

Q2 Solve the following inequalities:

a) $2x \geqslant 16$ ....

b) $4x > -20$ ....

c) $x + 2 > 5$ ....

d) $x - 3 \leqslant 10$ ....

e) $10x > -2$ ....

f) $5 + x \geqslant 12$ ....

g) $x/4 > 10$ ....

h) $x/3 \leqslant 1$ ....

i) $5x + 4 < 24$ ....

j) $5x + 7 \leqslant 32$ ....

k) $3x + 12 \leqslant 30$ ....

l) $2x - 7 \geqslant 8$ ....

Q3 There are 1,130 pupils in a school and no classes have more than 32 pupils. What is the least number of classrooms that could be used? Show this information as an inequality.

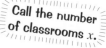

Call the number of classrooms x.

....

Q4 Jon has a budget of £10 to buy food for his dinner. He picks up a steak which costs £4.70 and a tin of sweetcorn which costs 90p. Potatoes cost 65p each. What is the greatest number of potatoes Jon can buy with the money he has left?

....

Number Patterns and Sequences

Q1 Draw the next two pictures in each pattern.
How many match sticks are used in each picture?

a)

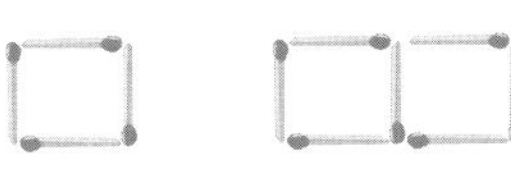

, , , ,

....

b)

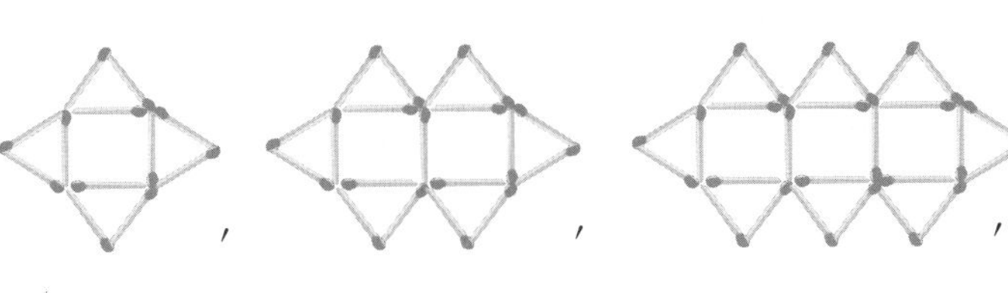

, , ,

....

Look for patterns in the numbers as well as pictures.

,

....

c)

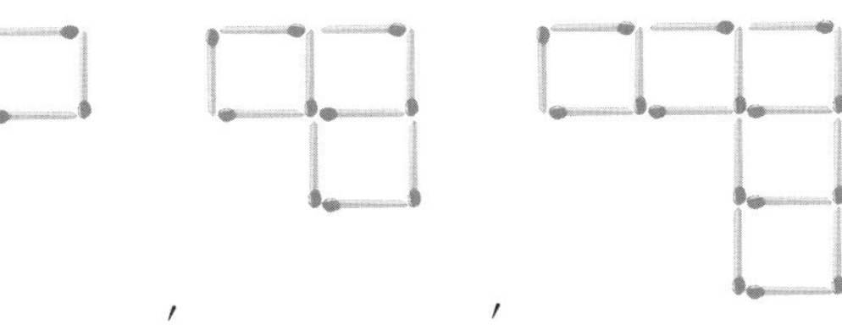

, , , ,

....

Q2 Look for the pattern and then fill in the next three lines. You're not allowed to use a calculator for this so you must spot the pattern.

a)

7 × 6 = 42

67 × 66 = 4422

667 × 666 = 444 222

6667 × 6666 = ...

66 667 × 66 666 = ...

666 667 × 666 666 = ...

b)

1 × 81 = 81

21 × 81 = 1701

321 × 81 = 26 001

4321 × 81 = 350 001

54 321 × 81 = ...

654 321 × 81 = ...

7 654 321 × 81 = ...

Number Patterns and Sequences

Once you've worked out the next numbers, go back and write down exactly what you did — that will be the rule you're after.

Q3 In each of the questions below, write down the next three numbers in the sequence and write the rule that you used.

a) 1, 3, 5, 7, ... , ... , ... Rule ....

b) 2, 4, 8, 16, ... , ... , ... Rule ....

c) 3, 30, 300, 3000, ... , ... , ... Rule ....

d) 3, 7, 11, 15, ... , ... , ... Rule ....

Q4 The letter $\boldsymbol{n}$ describes the position of a term in the sequence. For example, if $\boldsymbol{n} = 1$, that's the 1st term...if $\boldsymbol{n} = 10$ that's the 10th term and so on.
In the following, use the rule given to generate (or make) the first 4 terms.

a) $3\boldsymbol{n} + 1$ so if $\boldsymbol{n} = 1$ the 1st term is $\underline{(3 \times 1) + 1} = \underline{4}$

$\boldsymbol{n} = 2$ the 2nd term is .... = ...

$\boldsymbol{n} = 3$ .... = ...

$\boldsymbol{n} = 4$ .... = ...

b) $5\boldsymbol{n} - 2$, when $\boldsymbol{n} = 1, 2, 3, 4$ and 5
produces the sequence .... , ... , ... , ... , ...

c) $\boldsymbol{n}^2$, when $\boldsymbol{n} = 1, 2, 3, 4$, and 5
produces the sequence .... , ... , ... , ... , ...

Q5 Write down an expression for the nth term of the following sequences:

Remember the formula — $dn + (a - d)$.

a) 2, 4, 6, 8, ...

....

b) 1, 3, 5, 7, ...

....

c) 5, 8, 11, 14, ...

....

Q6 Mike has opened a bank account to save for a holiday. He opened the account with £20 and puts £15 into the account at the end of each week. So at the end of the first week the balance of the account is £35. What will the balance of the account be after:

a) 3 weeks? ....

b) 5 weeks? ....

c) Write down an expression that will let Mike work out how much he'll have in his account after any number of weeks he chooses.

....

Coordinates

Q1 On the grid plot the following points. Label the points A,B...
Join the points with straight lines as you plot them.

A(0,8) B(4,6) C(4.5,6.5) D(5,6) E(9,8) F(8,5.5) G(5,5) H(8,4) I(7.5,2) J(6,2) K(5,4) L(4.5,3.5) M(4,4) N(3,2) O(1.5,2) P(1,4) Q(4,5) R(1,5.5) S(0,8).

You should see the outline of an insect. What is it? ....

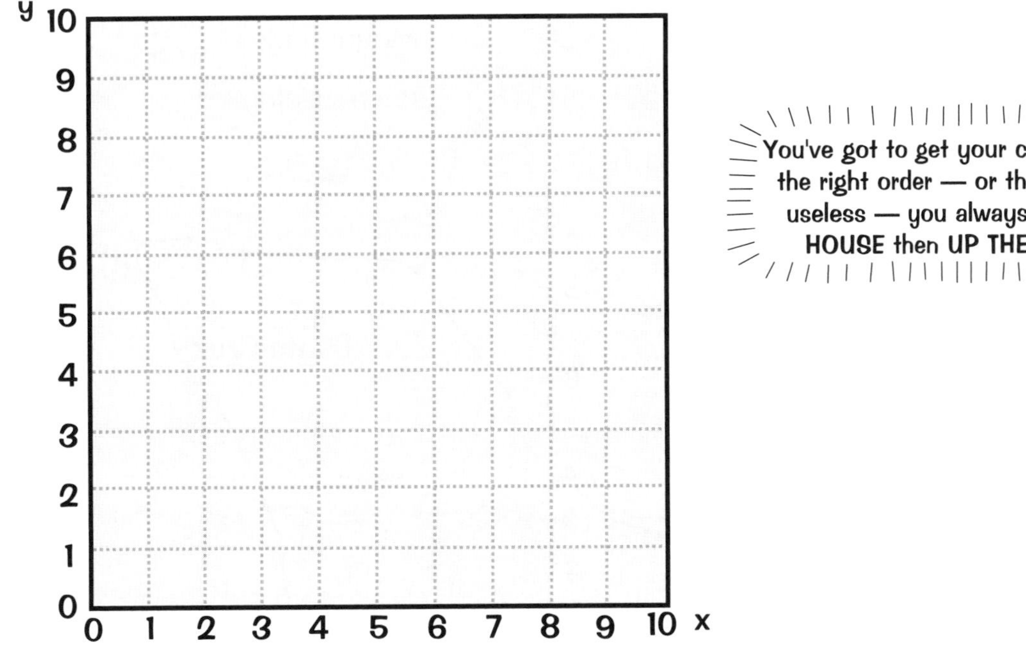

You've got to get your coordinates in the right order — or they're totally useless — you always go **IN THE HOUSE** then **UP THE STAIRS.**

Q2 Write down the letter which is by each of the following points.
The sentence it spells is the answer to question one.

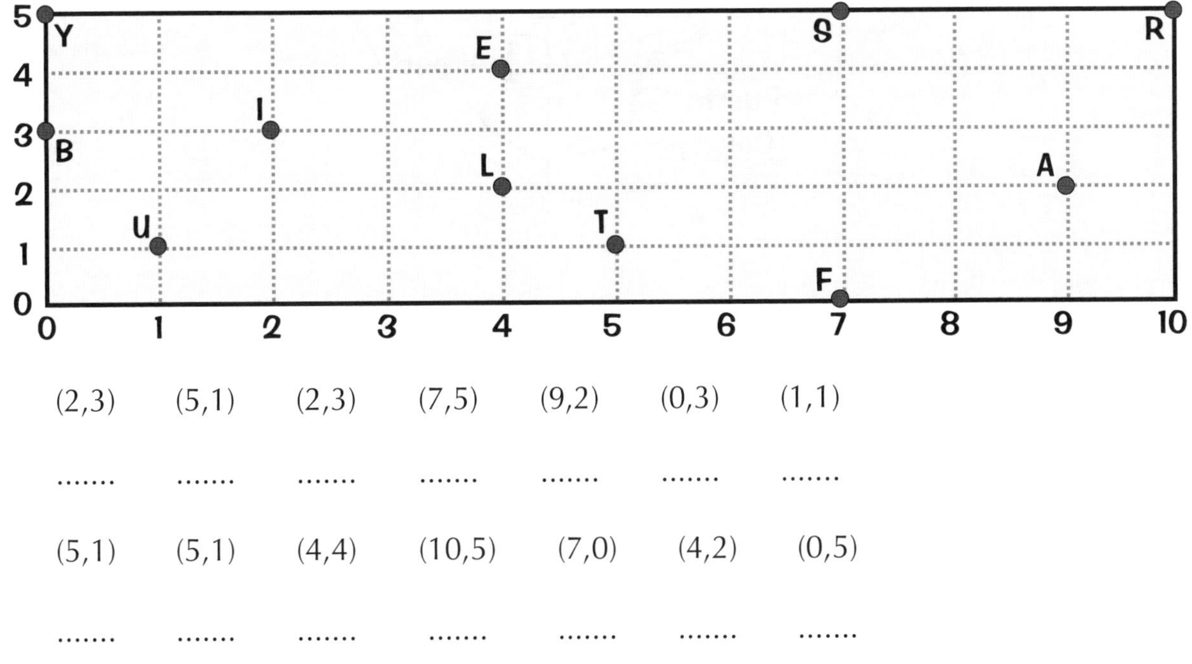

(2,3) (5,1) (2,3) (7,5) (9,2) (0,3) (1,1)

....

(5,1) (5,1) (4,4) (10,5) (7,0) (4,2) (0,5)

....

Coordinates

Remember — 1) x comes before y

2) x goes a-cross (get it) the page. (Ah, the old ones are the best...)

Q3 The map shows the island of Tenerife where the sun never stops shining...

a) Use the map to write down the coordinates of the following:

Airport (.... , ...) Mount Teide (.... , ...)

Santa Cruz (.... , ...) Puerto Colon (.... , ...)

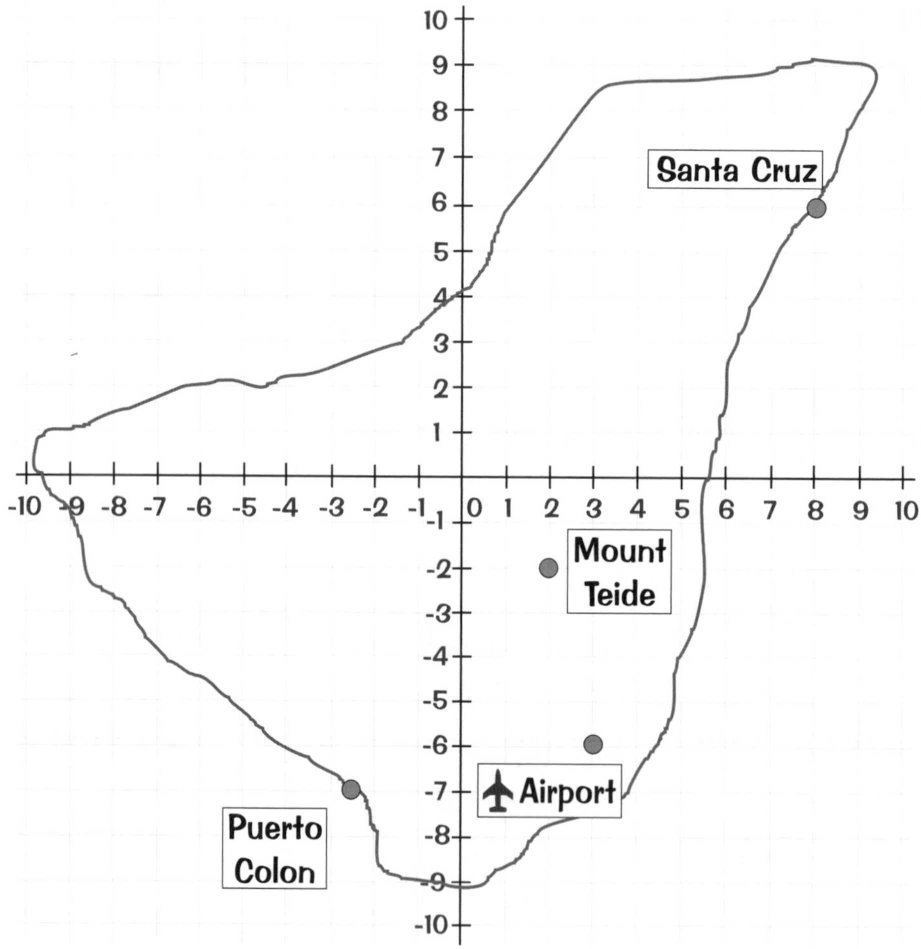

b) Use the coordinates given to mark the following holiday destinations on the map.
Las Americas (-4 , -6), El Medano (4 , -4), Icod (-6 , 2), Laguna (3 , 7), Taganana (9 , 9)

c) The cable car takes you to the top of Mount Teide. It starts at (3 , 1) and ends at (2 , -2).
Draw the cable car route on the map.

Plotting Straight Line Graphs

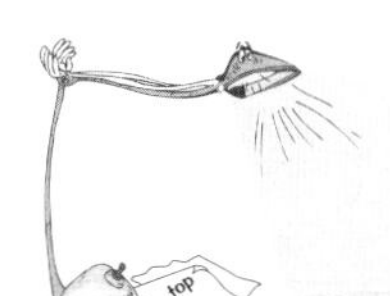

The <u>very first thing</u> you've got to do is work out a <u>table of values</u>.

Example: Draw the graph of $y = 3x - 1$ for values of x between 0 and 4.

1) First complete a table of values:

x	0	1	2	3	4
y	-1	2	5	8	11

decided by the question

worked out using y = 3x - 1

2) Draw the axes.
3) Plot the points.
4) Draw a straight line through the points.

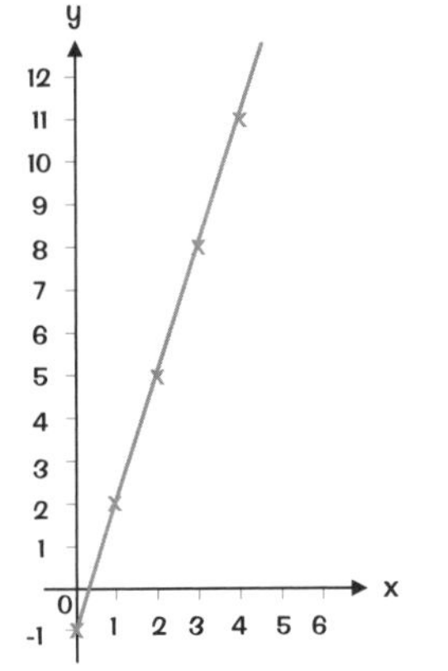

Q1 On the grid shown, draw axes with x from 0 to 8 and y from 0 to 14.

Q2 a) Complete the table of values, for $y = x + 2$.

x	0	1	2	3	4	5	6
y	2			5			

b) Use your <u>table of values</u> to draw the graph of $y = x + 2$ on the grid opposite.

Q3 a) Fill in the table for $y = 8 - x$, using values of x from 0 to 6.

x	0	1	2	3	4	5	6
y							

b) Draw the graph of $y = 8 - x$ on the grid opposite.

Q4 A rough way of changing temperatures from Celsius to Fahrenheit is to 'double it and add 30' ($y = 2x + 30$).

a) Fill in the table below to change the temperatures in Celsius (x) to Fahrenheit (y).

x	0	5	10	15	20	25	30
y	30				70		

b) Use your <u>table of values</u> to draw a temperature conversion graph. Draw and label the axes with x from 0 to 30, and y from 0 to 100.

Gradients of Lines

The gradient of a line is just a measure of the slope. The box below tells you everything you need to know about it...

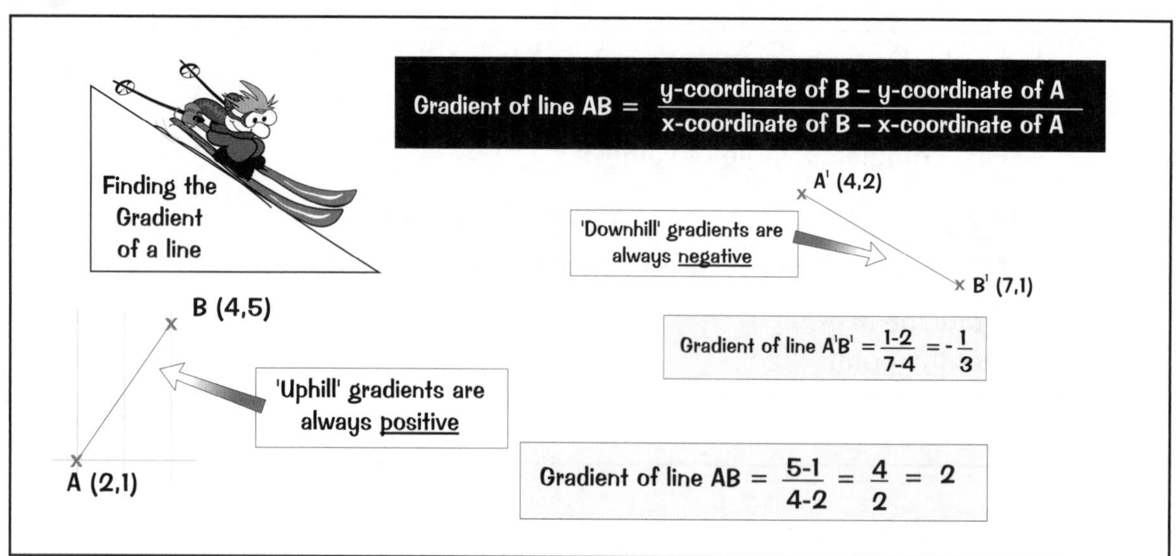

Q1 Draw axes with x from -9 to 9 and y from -12 to 12. On this set of axes join each <u>pair of points</u> and work out the <u>gradient</u> of the line.

A is (1, 1), B is (2, 4)

gradient of AB = ...

C is (5, 5), D is (7, 0)

gradient of CD = ...

E is (-7, 7), F is (-2, 10)

gradient of EF = ...

G is (-6, 2), H is (-3,-4)

gradient of GH = ...

I is (-8,-9), J is (-3,-6)

gradient of IJ = ...

Interpreting Graphs

Q1 A group of friends are on holiday in France. The distances to different places are given in kilometres in the guidebook, but there is a handy conversion graph to help them get a feel for the distances in miles.

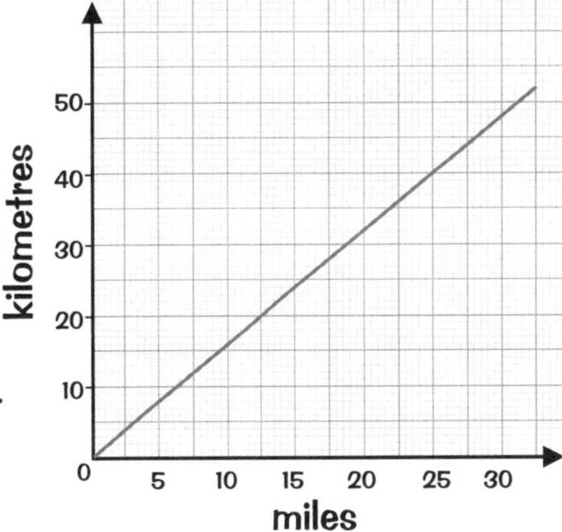

a) To the nearest mile, how far away is:

the beach — 10 km ....

the airport — 35 km ....

the shopping centre — 27 km ....

b) The group are from a town 23 miles outside London. A French friend asks what this is in kilometres. Use the graph to find the answer to the nearest km.

....

Q2 Match the following graphs with the statements below:

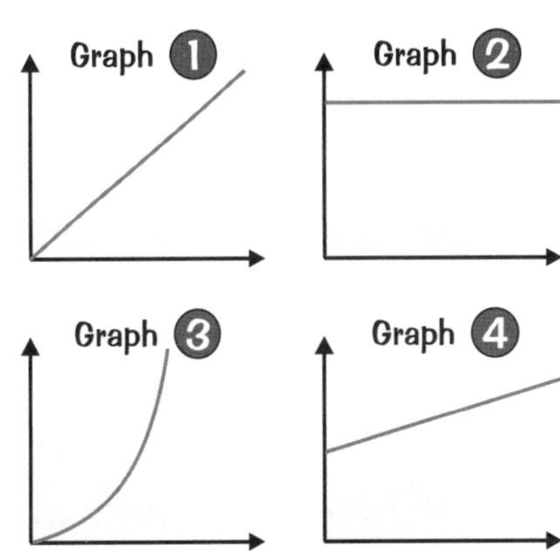

a) The cost of hiring a plumber per hour including a fixed call-out fee. ....

b) Exchange rate between Euros and American Dollars ....

c) Speed against time for a car travelling at constant speed. ....

d) The area of a circle as the radius increases. ....

Q3 In her science lesson, Ellie pours water into different shaped containers at a constant rate, then plots graphs of the depth of water (d) against time (t) taken to fill the container.

At the end of the lesson she realises she hasn't labelled her graphs. Which graph matches each container? Write in the letters below.

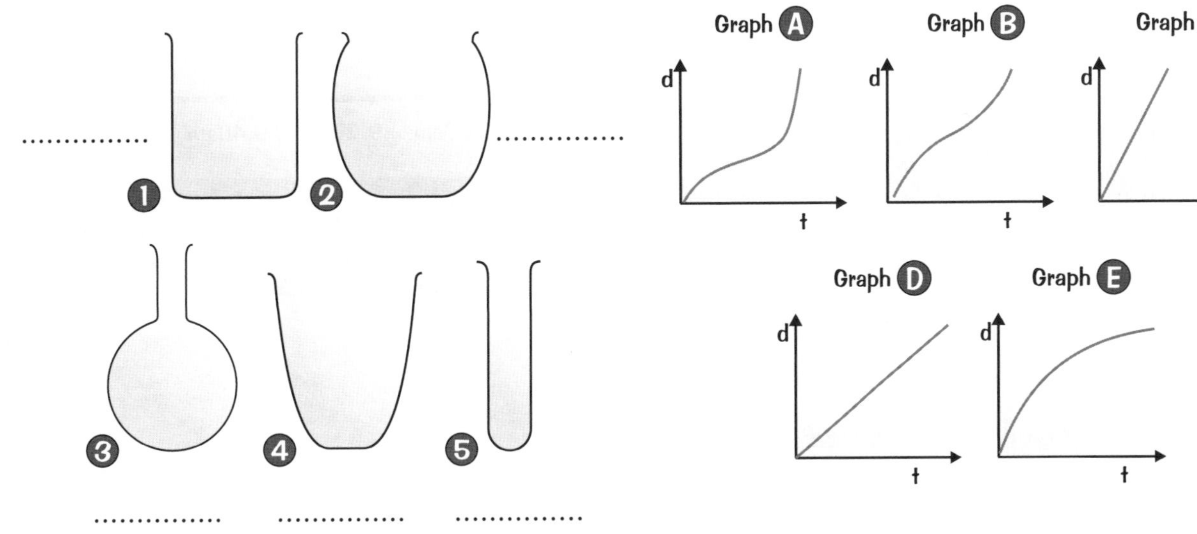

Travel Graphs

Q1 The graph shows Nicola's car journey from her house to Alan's house, picking up Robbie on the way.

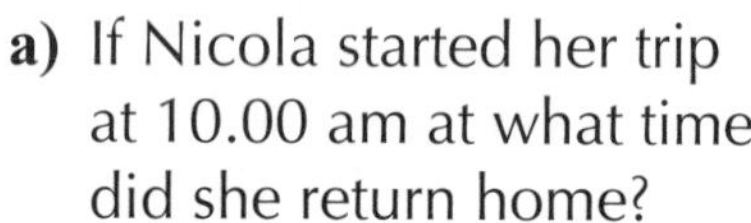

a) If Nicola started her trip at 10.00 am at what time did she return home? ....

b) How far is Robbie's house from Nicola's? ....

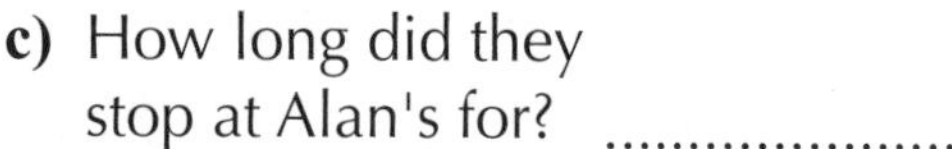

c) How long did they stop at Alan's for? ....

d) During which section was the speed greatest? ....

e) How long did the return journey take? ....

f) What was the speed of the car during section E? ....

You can work out where the houses are by looking for the flat parts of the graph — the bits where Nicola stops. There are two flat parts here — one for Robbie's house and one for Alan's.

Q2 Marcus competes in a 10 km race. All the runners are given a small device to wear which records the time they pass through certain checkpoints. Later, Marcus gets a graph of his performance during the race, shown below.

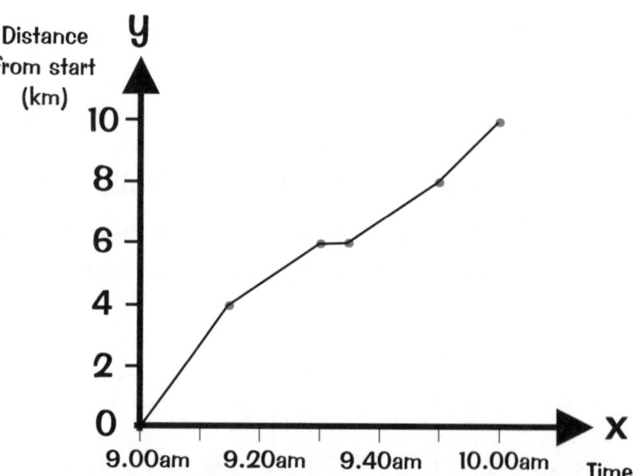

a) Between what times was Marcus running the <u>fastest</u>?

....

b) Calculate his <u>fastest speed</u> in km/hr.

....

c) What time did Marcus <u>stop</u> for a drink?

....

d) For <u>how long</u> did he stop?

....

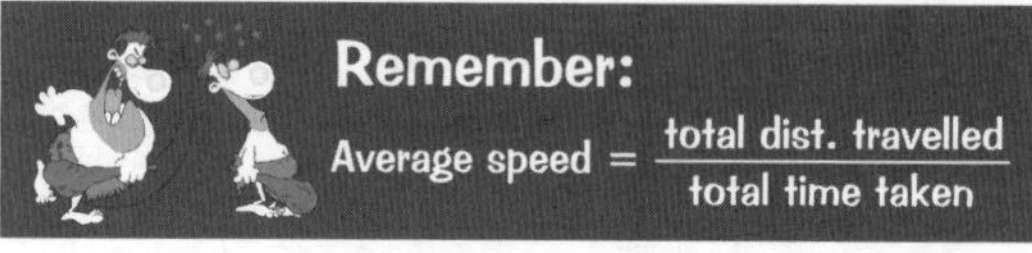

e) How long did it take Marcus to <u>complete</u> the 10 km run (in hours)?

....

f) What was the <u>average speed</u> for his entire run?

....

Trial and Improvement

Q1 Use the trial and improvement method to solve the equation $x^3 = 50$.
Give your answer to one decimal place. Two trials have been done for you.

Try $x = 3$ $x^3 = 27$ (too small)
Try $x = 4$ $x^3 = 64$ (too big)

....

Q2 Use the trial and improvement method to solve these equations.
Give your answers to one decimal place.

a) $x^2 + x = 80$

....

b) $x^3 - x = 100$

....

Show all the numbers you've tried, not just your final answer... or you'll be chucking away easy marks.

Quadratic Graphs

If an expression has an x^2 term in it, it's quadratic. The graphs you get from quadratic expressions are always curves with a certain shape...

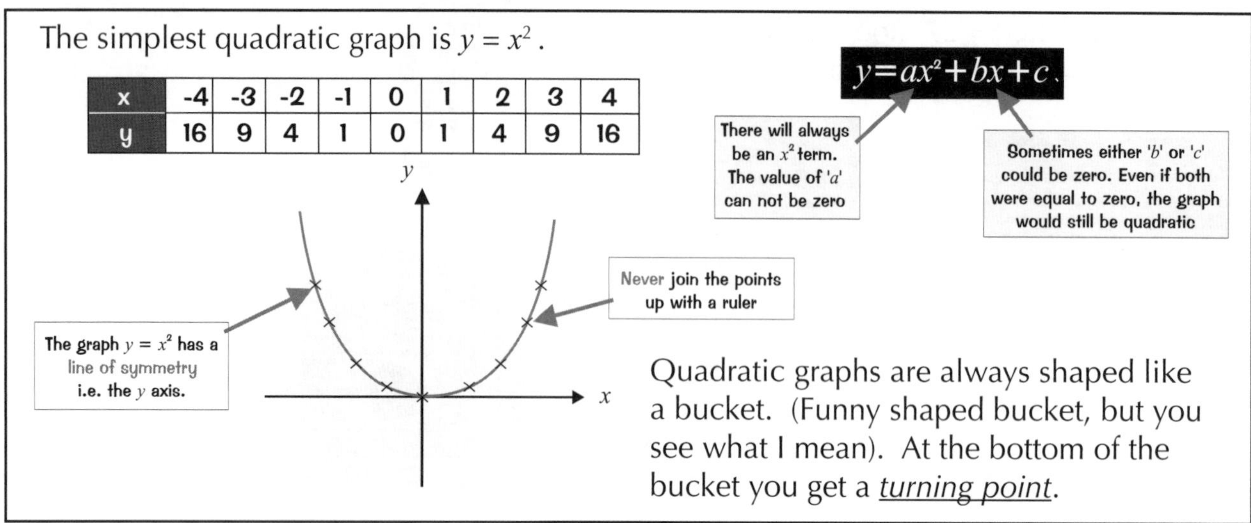

The simplest quadratic graph is $y = x^2$.
x | -4 | -3 | -2 | -1 | 0 | 1 | 2 | 3 | 4
y | 16 | 9 | 4 | 1 | 0 | 1 | 4 | 9 | 16
$y=ax^2+bx+c$.
There will always be an x^2 term. The value of 'a' can not be zero
Sometimes either 'b' or 'c' could be zero. Even if both were equal to zero, the graph would still be quadratic
Never join the points up with a ruler
The graph $y = x^2$ has a line of symmetry i.e. the y axis.
y
x

Quadratic graphs are always shaped like a bucket. (Funny shaped bucket, but you see what I mean). At the bottom of the bucket you get a turning point.

You'll need some graph paper for the questions on this page.

Q1 Complete this table of values for the quadratic graph $y = 2x^2$.

a) On your graph paper, draw axes with x from -4 to 4 and y from 0 to 32.

b) Plot these 9 points and join them with a smooth curve.

x	-4	-3	-2	-1	0	1	2	3	4
$y=2x^2$	32	18					8		

Remember to square first then x 2

Q2 Complete this table of values for the graph $y = x^2 + x$.

x	-4	-3	-2	-1	0	1	2	3	4
x^2	16	9					4		
$y=x^2+x$	12					2			

By putting more steps in your table of values, the arithmetic is easier

a) Draw axes with x from -4 to 4 and y from 0 to 20.

b) Plot the points and join them with a smooth curve.

c) Draw the line of symmetry for the quadratic graph $y = x^2 + x$, and label it.

d) Use your graph to find the two approximate values of x when $y = 0$ for the equation $y = x^2 + x$.

$x =$ .... or ....

Just find the x-values where the graph crosses the x-axis (i.e. where $y = 0$).

Symmetry

To work out if you've got a line of symmetry, just imagine you're folding the shape in half. If the sides will fold exactly together, then hey presto, it's symmetrical about the fold.

Q1 These shapes have more than one line of symmetry.
Draw the lines of symmetry using dotted lines.

a) 

b) 

c) 

Q2 Some of the letters of the alphabet have lines of symmetry.
Draw the lines of symmetry using dotted lines.

A B C D E F G H I J K L M

N O P Q R S T U V W X Y Z

Q3 Complete the following diagrams so that they have rotational symmetry about centre C of the order stated:

a) order 2

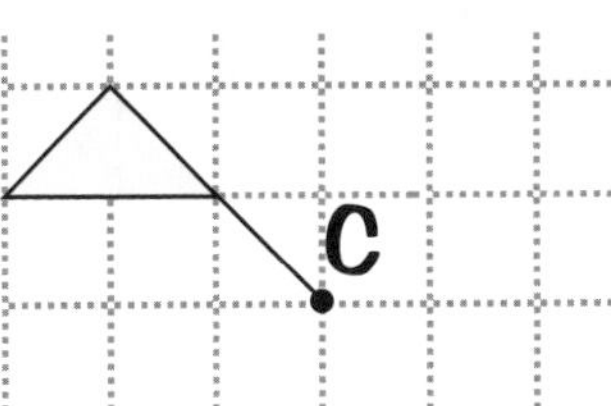

b) order 4

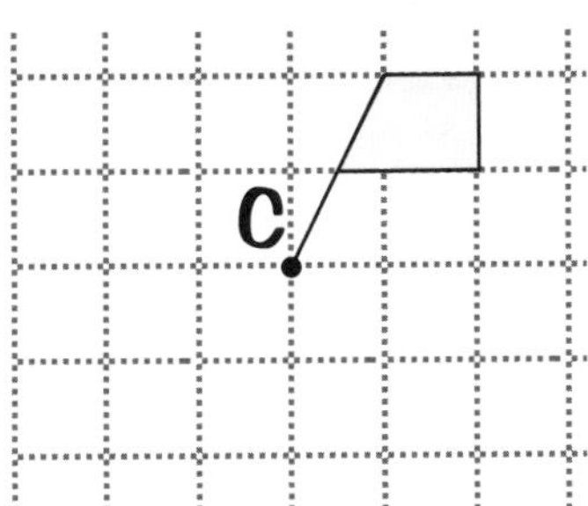

c) order 3

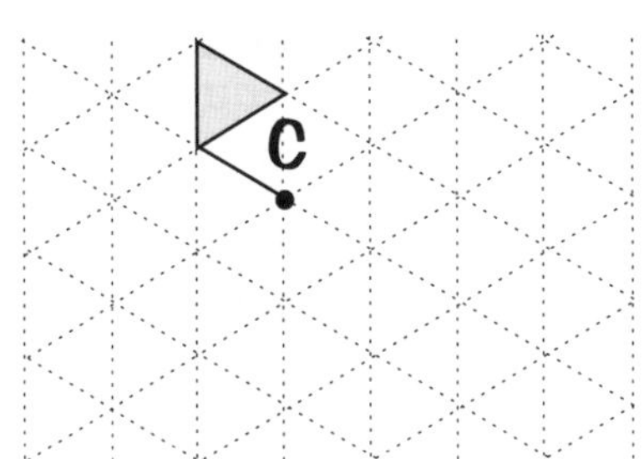

Symmetry

Q4 The diagram shows part of a shape with one line of symmetry. Complete the drawing of the shape, using the given line of symmetry.

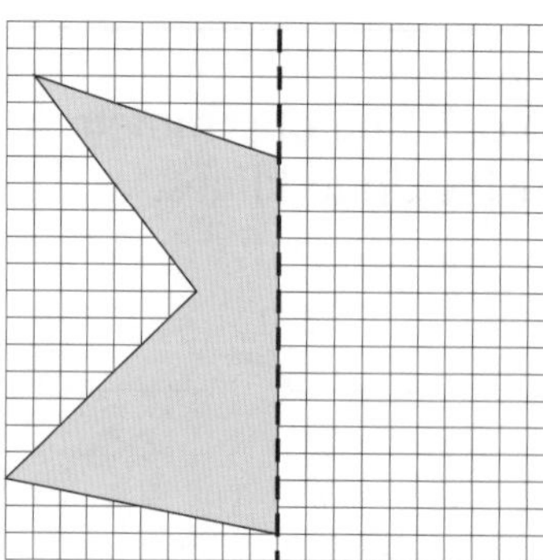

Q5 Write down the number of lines of symmetry, and the order of rotational symmetry of each of the following shapes:

a) 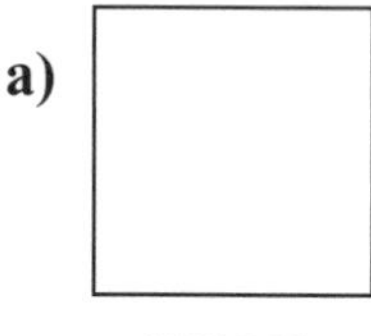
square

b) 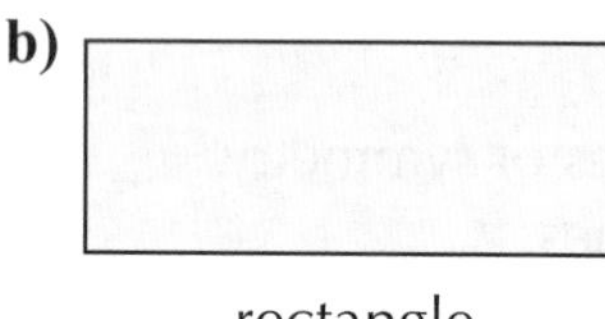
rectangle

c) 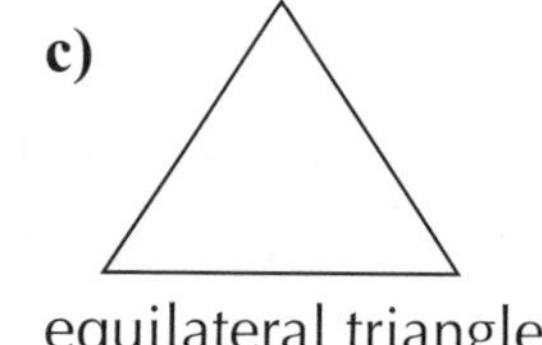
equilateral triangle

d) 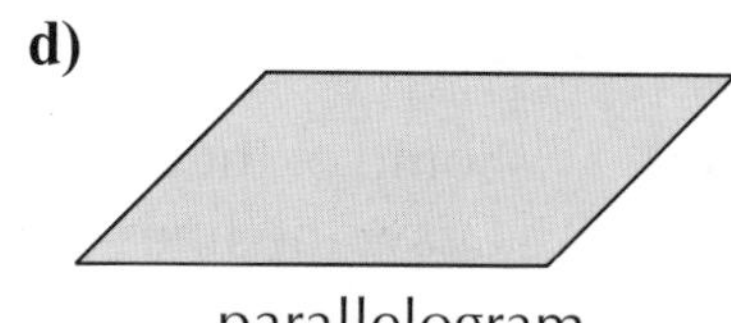
parallelogram

Q6 The polygon shown here is a hexagon.

a) Draw in all its lines of symmetry.
b) State the order of rotational symmetry.

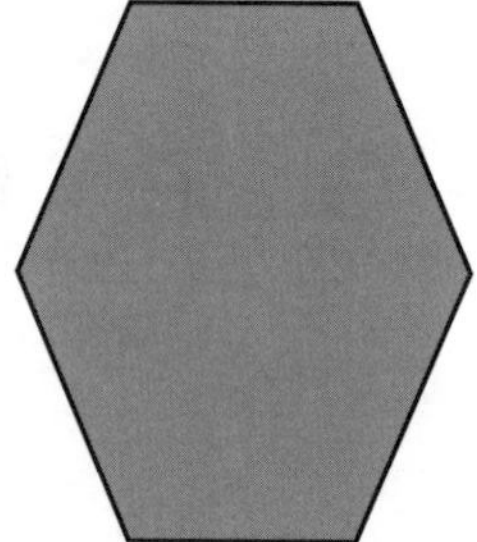

Order of rotational symmetry is ....

Q7 Complete the following table:

Name	Sides	Lines of Symmetry	Order of Rotational Symmetry
Equilateral Triangle			
Square		4	
Regular Pentagon			
Regular Hexagon	6		
Regular Heptagon	7		
Regular Octagon			8
Regular Decagon	10		

Families of Triangle

Q1 Fill in the gaps in these sentences.

a) An isosceles triangle has .... equal sides and .... equal angles.

b) A triangle with all its sides equal and all its angles equal is called an .... triangle.

c) A scalene triangle has .... equal sides and .... equal angles.

d) A triangle with one right-angle is called a .... triangle.

Q2 By joining dots, draw four different isosceles triangles — one in each box.

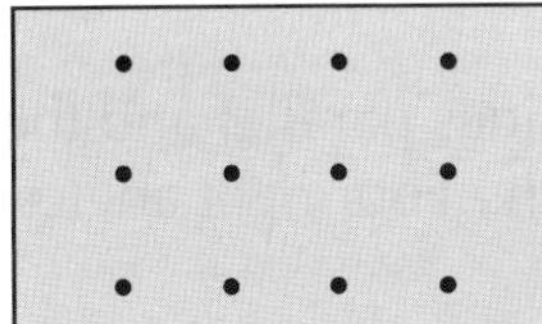
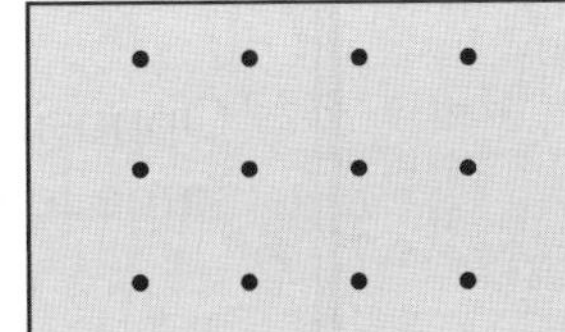
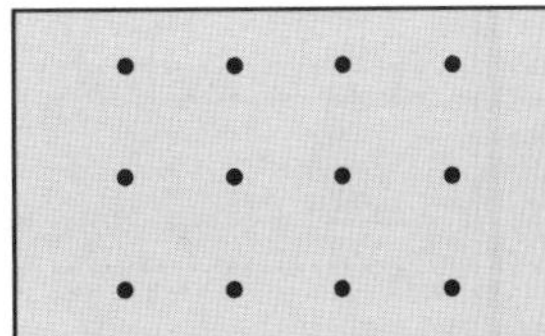
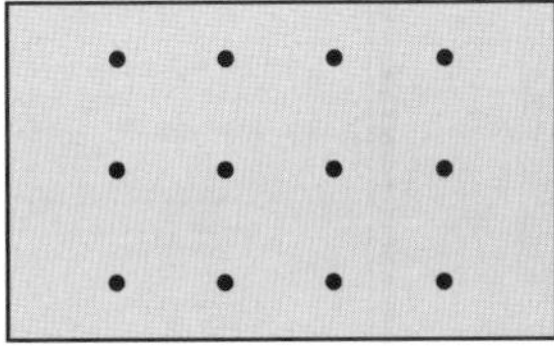

Q3 Using three different coloured pencils:
Find an equilateral triangle and shade it in.
Using a different colour, shade in two different right angled triangles.
With your last colour shade in two different scalene triangles.

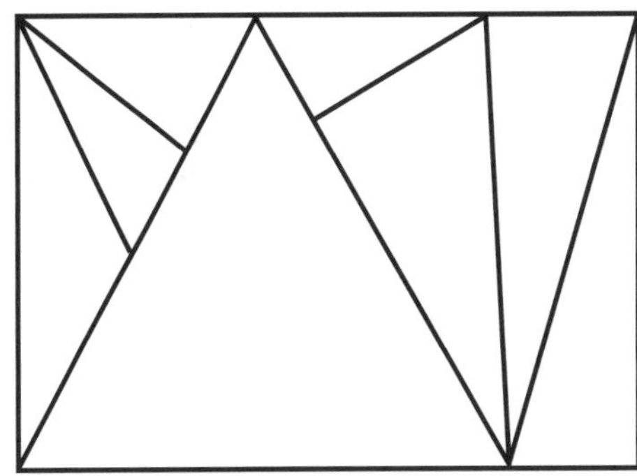

Q4 How many triangles are there in this diagram? ....

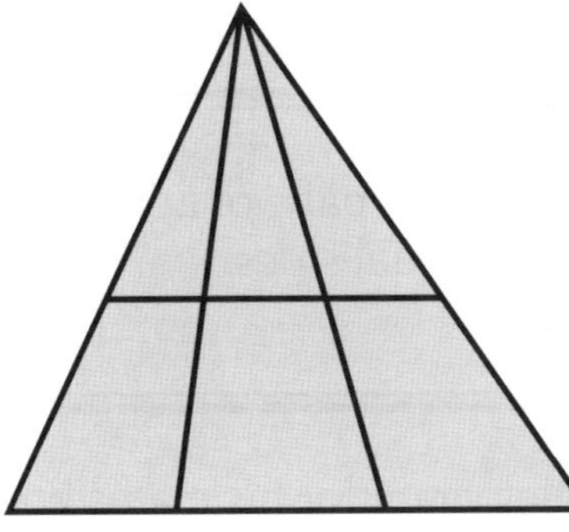

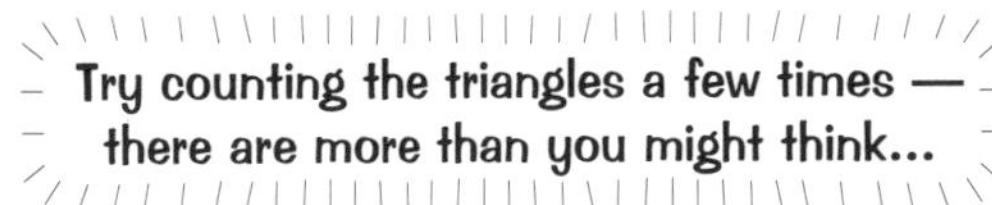

Try counting the triangles a few times —
there are more than you might think...

What sort of triangles are they?

....

There are only 4 types of triangle, so make sure you know them all — think of all those nice fat juicy marks...

Quadrilaterals

Here's a few easy marks for you — all you've got to do is remember the shapes and a few facts about them... it's a waste of marks not to bother.

Q1 Fill in the blanks in the table.

NAME	DRAWING	DESCRIPTION
Square		Sides of equal length. Opposite sides parallel. Four right angles.
....	....	Opposite sides parallel and the same length. Four right angles.
....		Opposite sides are .... and equal. Opposite angles are equal.
Trapezium		Only .... sides are parallel.
Rhombus		A parallelogram but with all sides ....
Kite	....	Two pairs of adjacent equal sides.

Q2 Stephen is measuring the angles inside a parallelogram for his maths homework. To save time, he measures just one and works out what the other angles must be from this. If the angle he measures is 52°, what are the other three angles?

Hint: the angles in a parallelogram add up to 360°.

.........., ..., ...,

Regular Polygons

You need to remember these two formulas for polygons:

Sum of Exterior angles = 360°

and Sum of Interior angles = (n – 2) × 180°

(n is the number of sides)

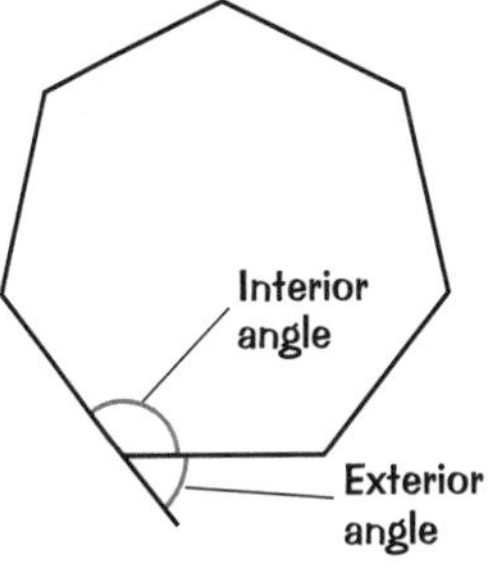

Q1 Here is a regular octagon:

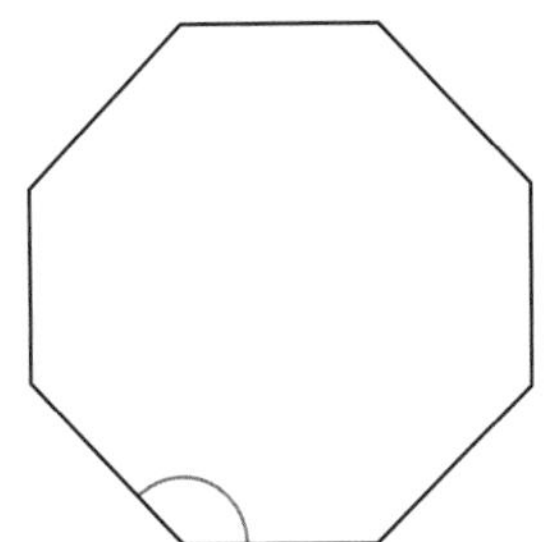

a) What is the total of its eight interior angles? ....

b) What is the size of the marked angle? ....

Here is a regular pentagon:

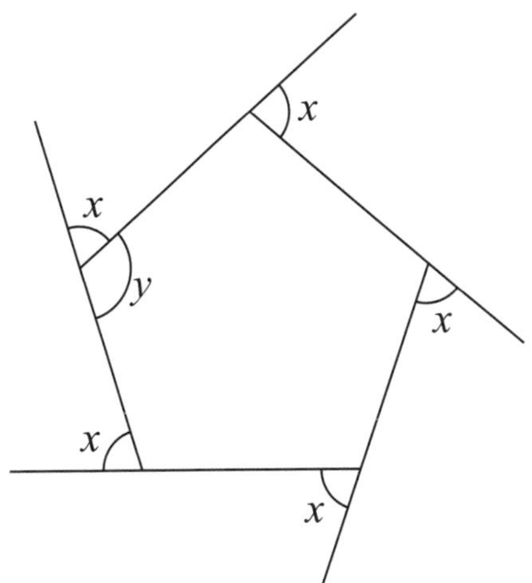

c) The five angles marked x are its exterior angles. What do they add up to?

....

d) Work out the value of x.

....

e) Use your answer from part **d)** to work out the value of angle y.

....

Q2 Charlotte is making a template for a patchwork quilt made from regular hexagons. She needs to measure and cut her template accurately before she uses it to cut the fabric.

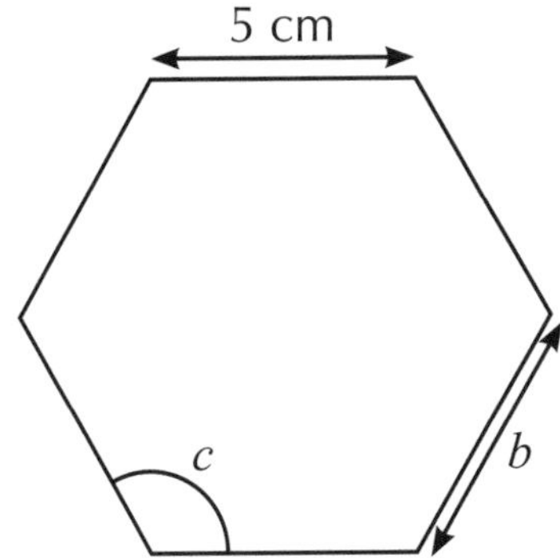

a) How long should she make side b? ....

b) What is the size of angle c? ....

c) Charlotte decides to make each hexagon out of six triangles instead. What type of triangle should she use?

....

Perimeters

What you've gotta do with these is add up all the sides to get the perimeter. If there's no drawing, do one yourself — then you won't forget any of the sides.

Q1 Work out the perimeters of the following shapes :

a)

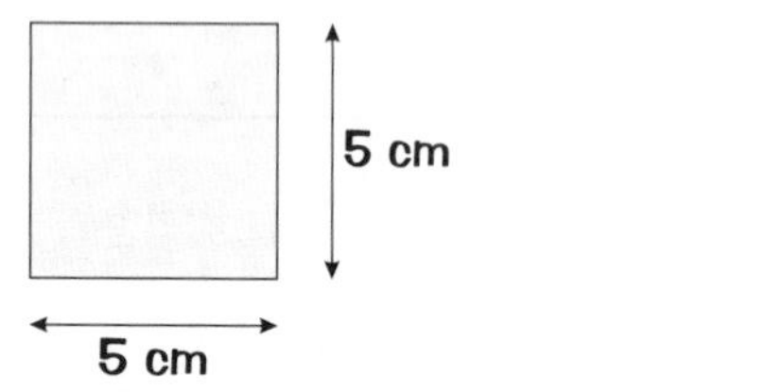

Square Perimeter = ... cm

b)

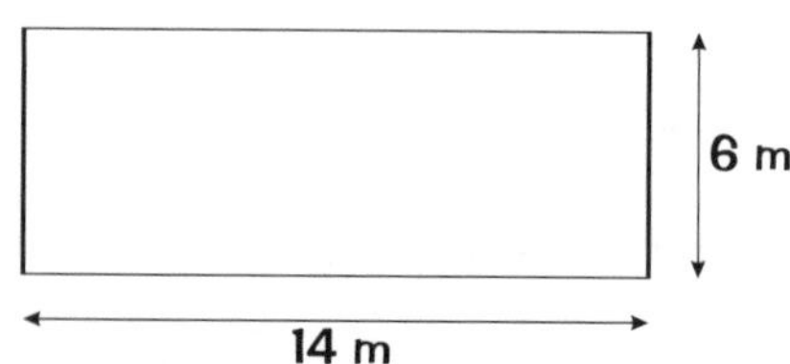

Rectangle Perimeter =

(2 × ......) + (2 × ......) = ... m

c)

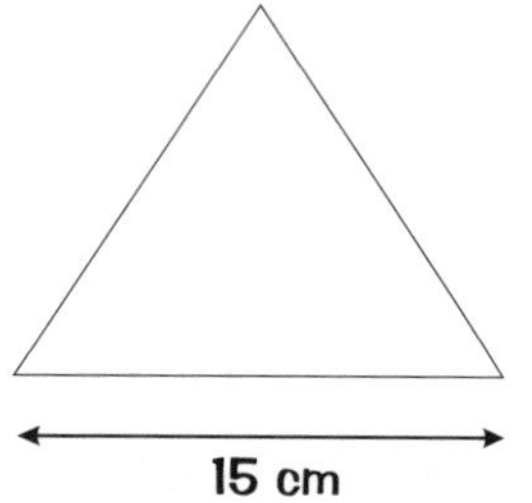

Equilateral Triangle

Perimeter = 3 × ...... = ... cm

d)

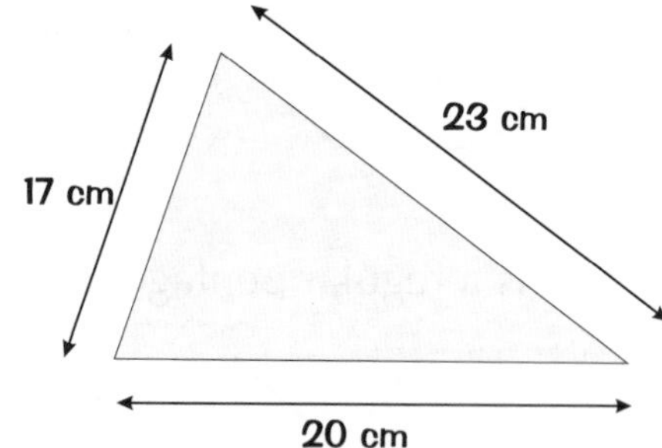

Triangle Perimeter = ... + ... + ...

= .... cm

e)

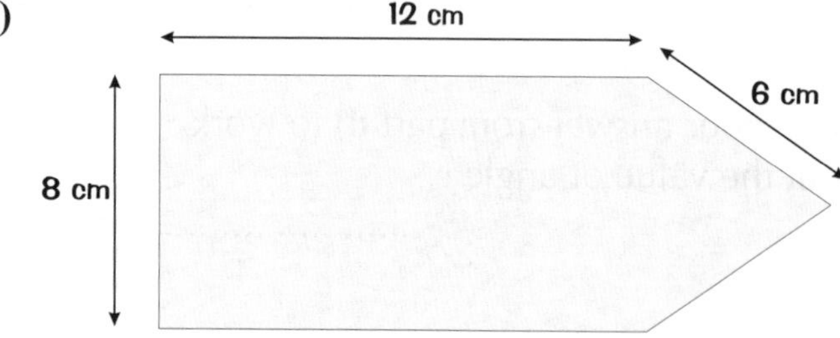

Symmetrical Five Sided Shape

Perimeter = ... + ... + ... + ... + ...

= .... cm

f)

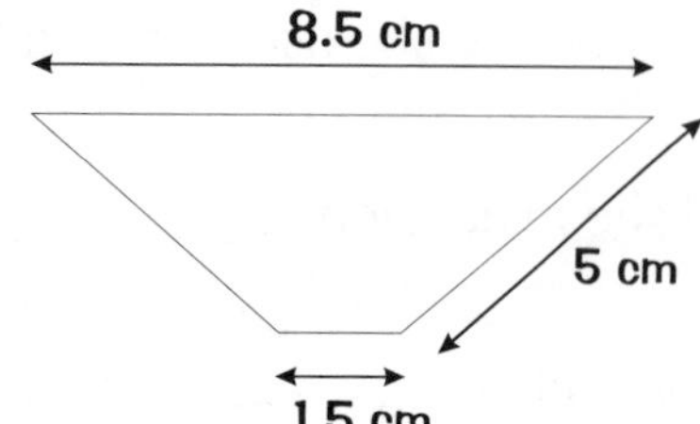

Symmetrical Four Sided Shape

Perimeter = ... + ... + ... + ... = ... cm

Q2 A square garden has sides of length 10 m.
How much fencing is needed to go around it? .... m

Perimeters

Q3 Find the perimeter of these shapes (you may need to work out some of the lengths):

a)

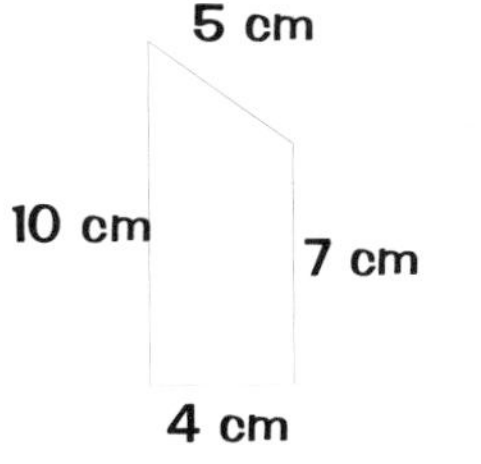

Perimeter ....

b)

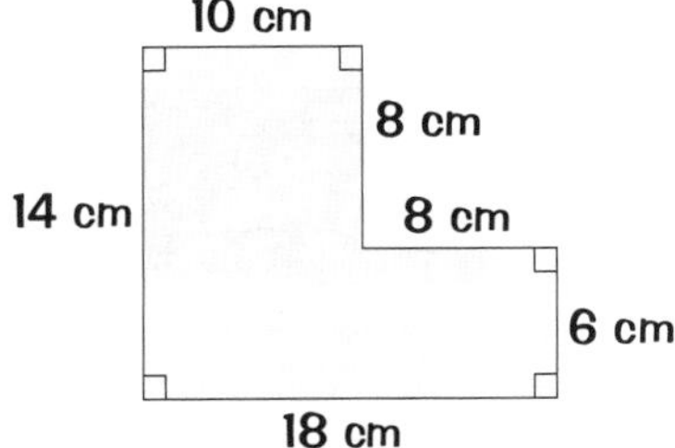

Perimeter ....

c)

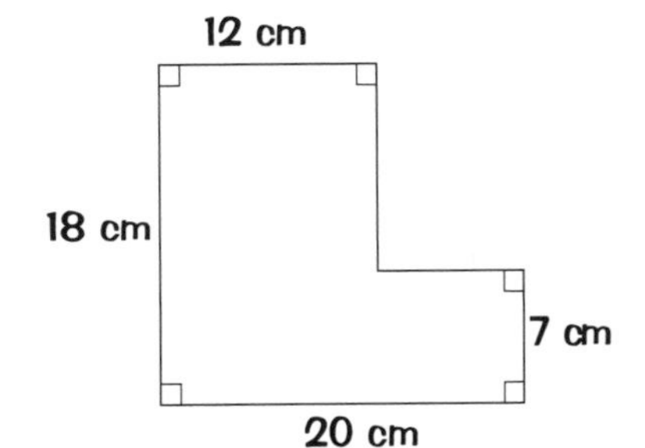

Perimeter ....

d)

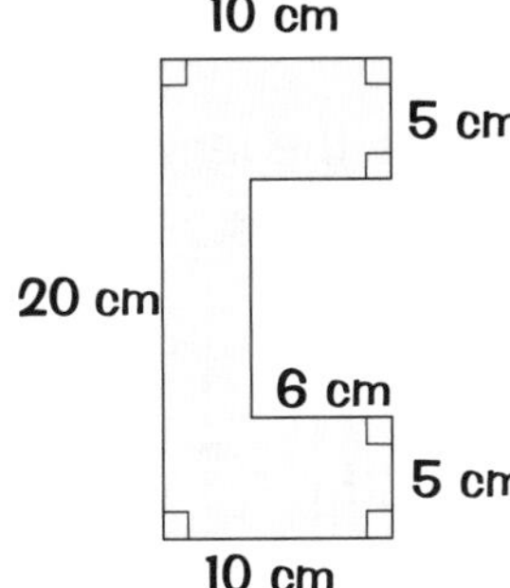

Perimeter ....

Q4 Isaac is putting a paper border across all of his living room walls. He has made a plan of the room with dimensions, shown here.

8 m
9 m
10 m
12 m
7 m
15 m

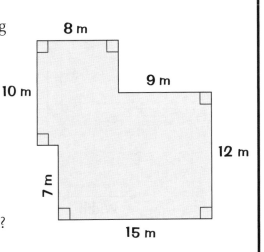

a) What is the total length of border he would need to go around the whole room?

....

b) The border comes in rolls of 5 m. How many would Isaac need to buy for his room?

....

Areas of Rectangles

For rectangles and squares, working out the area is a piece of pie — it's just LENGTH TIMES WIDTH. Nowt more to it.

AREA = LENGTH × WIDTH

Q1 Calculate the areas of the following rectangles:

a) Length = 10 cm, Width = 4 cm, Area = ... × ..... = ... cm^2.

b) Length = 55 cm, Width = 19 cm, Area = ... cm^2.

c) Length = 155 m, Width = 28 m, Area = ... m^2.

d) Length = 3.7 km, Width = 1.5 km, Area = ... km^2.

Q2 Measure the lengths and widths of each of these rectangles, then calculate the area.

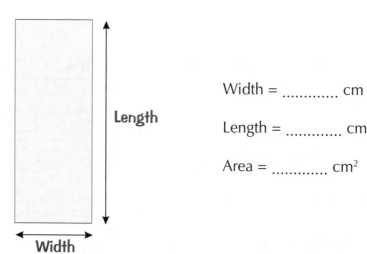

a)

Width

Length

Width = ... cm

Length = ... cm

Area = ... cm^2

b)

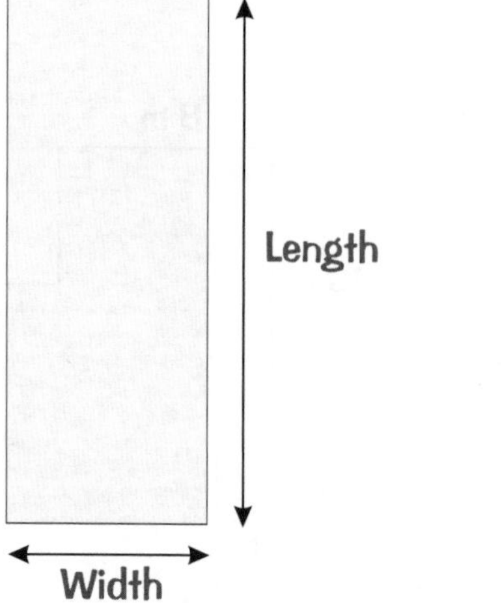

Width = ... cm

Length = ... cm

Area = ... cm^2

Q3 Sabrina is buying carpet for two rectangular rooms in her house.
One room is 4.8 m long and 3.9 m wide.
The other room is 4.2 m long and 3.1 m wide.
How many square metres of carpet does she need to buy?

....

Areas of Triangles

Triangles aren't much different, but <u>remember</u> to TIMES BY THE ½.

Area = ½ (Base × Height)

Q1 Calculate the areas of the following triangles:

a) Base = 12 cm, Height = 9 cm, Area = ½ (....... × ....) = ... cm².

b) Base = 5 cm, Height = 3 cm, Area = ... cm².

c) Base = 25 m, Height = 7 m, Area = ... m².

d) Base = 1.6 m, Height = 6.4 m, Area = ... m².

e) Base = 700 cm, Height = 350 cm, Area = ... cm².

Q2 Measure the base and height of each of these triangles, then calculate the area.

a)

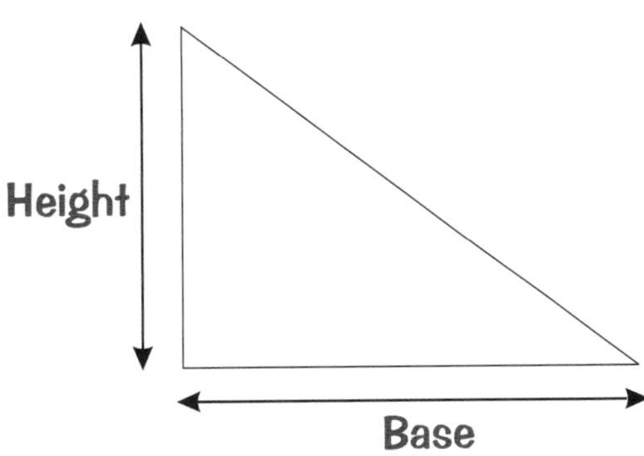

Height = ... cm

Base = ... cm

Area = ... cm²

b)

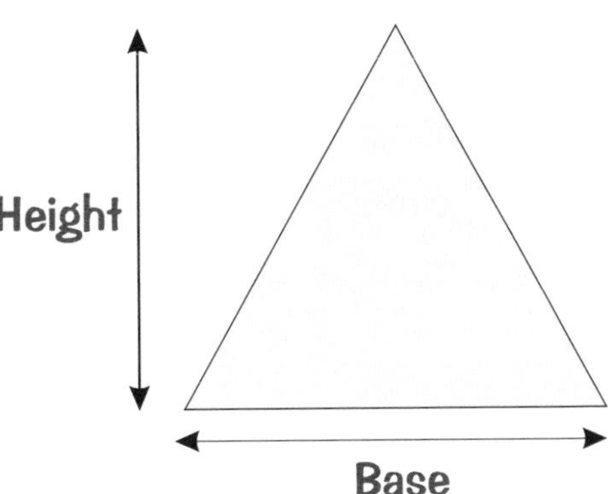

Height = ... cm

Base = ... cm

Area = ... cm²

Q3 Peter is making a kite. He has made a frame from wooden poles. What is the minimum area of fabric he needs to buy to cover the kite?

....

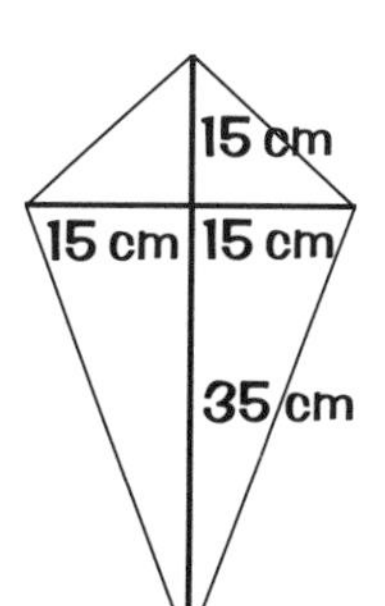

Think of the fabric he needs as four triangles sewn together.

Composite Areas

Bit more tricky, these... but look — they're all just rectangles and triangles. Work out each bit separately, then add the areas together — easy.

Q1 Calculate the areas of these composite shapes...

a)

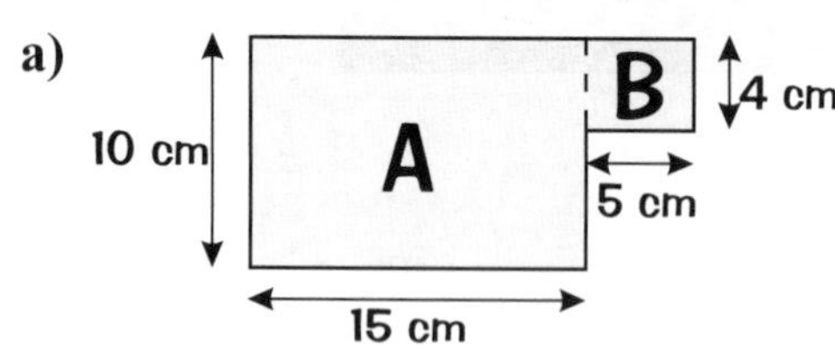

Shape A: length = ... width = ...

Area = ...× ....= ....cm^2

Shape B: length = ... width = ...

Area = ... × .... = ... cm^2

Total (area A + area B) = ... + ... = ... cm^2

b) Shape A (rectangle): .... × .... = ... cm^2

Shape B (triangle): ½ (base x height)

Base = ... cm , Height = ... cm

Area = ½ (........ × ....) = ... cm^2

Total Area = ... + ... = ... cm^2

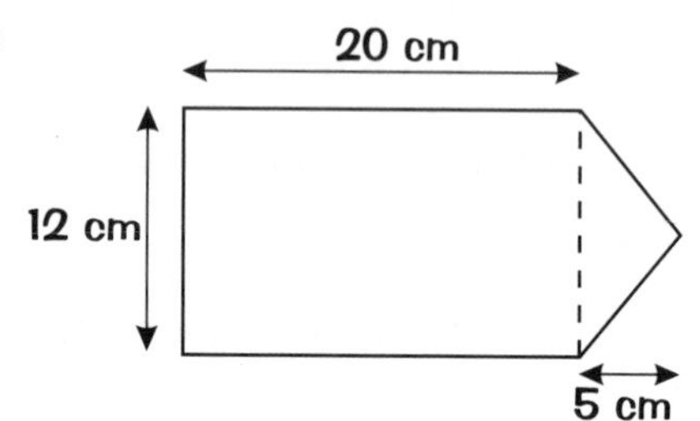

c)

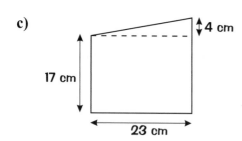

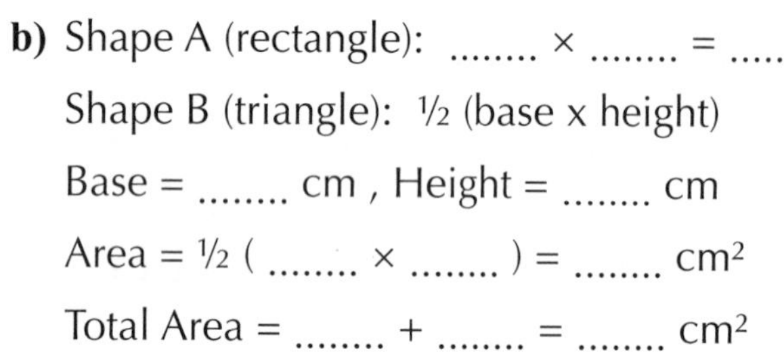

Shape A (rectangle): .... × .... = ... cm^2

Shape B (triangle): ½ (........ × ....) = ... cm^2

Total Area = ... + ... = ... cm^2

d) Draw a dotted line to divide this shape.

Shape A (rectangle): .... × .... = ... cm^2

Shape B (triangle): ½(........ × ....) = ... cm^2

Total Area = ... + ... = ... cm^2

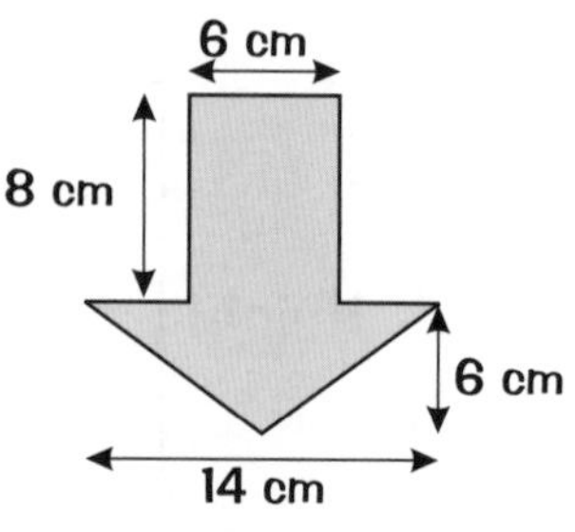

Q2 Tariq is making a gravelled patio in the shape shown. He can buy gravel in bags that cover 1 m^2 each. How many bags of gravel does he need to buy?

....

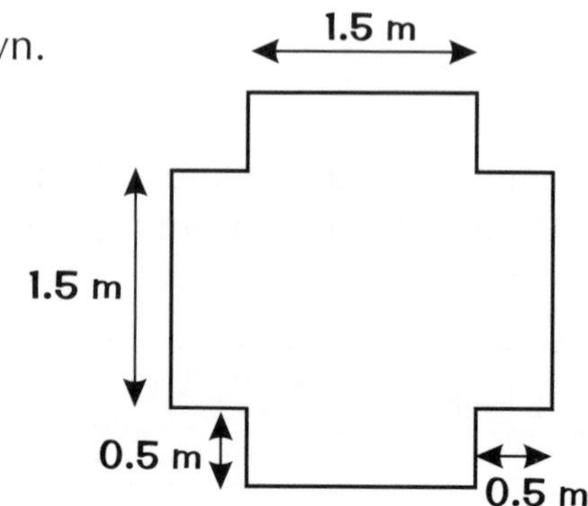

More Areas

Q1 This parallelogram has an area of 4773 mm².

How long is its base?

....

Remember, a parallelogram is just a sloping rectangle — so area = base × vertical height.

Q2 A simple tent is to be made in the shape of a triangular prism. The dimensions are shown in the diagram.

a) The two end faces are isosceles triangles. Find their areas.

....

....

b) The two sides are rectangles. Find their areas.

....

c) The groundsheet is also a rectangle. Find its area.

....

d) How much material is required to make this tent?

....

Q3 A lawn is to be made 48 m².

a) If its width is 5 m, how long is it?

....

b) Rolls of turf are 50 cm wide and 11 m long. How many rolls need to be ordered to grass the lawn?

....

Start by finding the area of 1 roll. Then work out how many rolls fit into the area of the lawn.

Circles

Don't worry about that π bit — it just stands for the number 3.14159... Sometimes you'll be told to round it off to 3 or 3.14. If not, just use the π button on your calculator.

Q1 Draw a circle with radius 3 cm.
On your circle label the circumference,
a radius and a diameter.

Q2 Calculate the circumference of these circles.
Take π to be 3.14.

a)

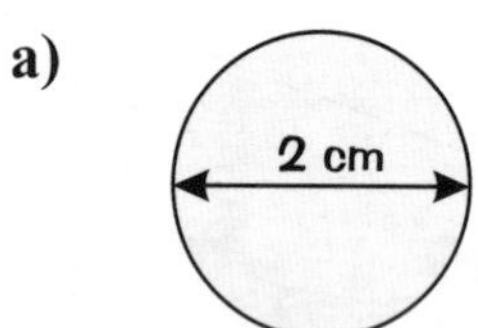

Circumference = π × diameter = ...

b) Circumference = π × diameter = ...

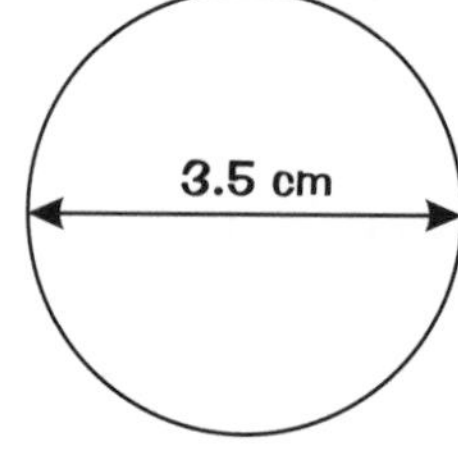

c)

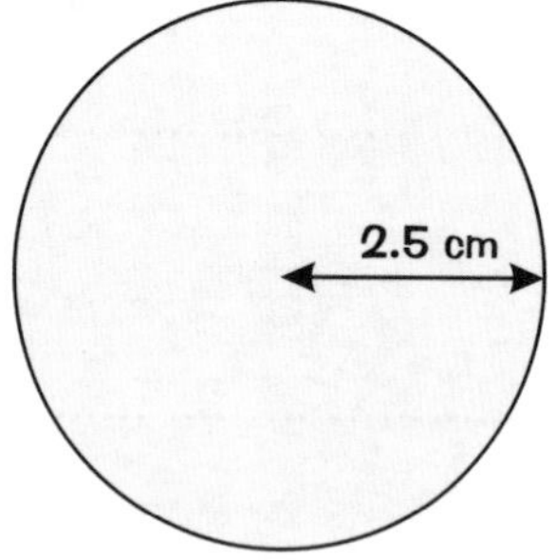

Remember to work out the diameter first.

Diameter = radius × 2 = ...

Circumference = ...

d) A circle of radius 3.5 cm. ....

Q3 A wheel has a diameter of 0.6 m. How far does it travel in one complete turn?

....

Q4 Sarah wants to decorate a plant pot by gluing a ribbon all the way around the top. The radius of the top of the pot is 4.5 cm.
What is the shortest piece of ribbon that Sarah could use for this?

....

Q5 The circumference of a circle is 195 cm.
Calculate its diameter correct to 3 sig. figs.

Diameter = ...

Circles

Q6 Calculate the area of each circle.
Write your answers to **a)** and **b)** as multiples of π.

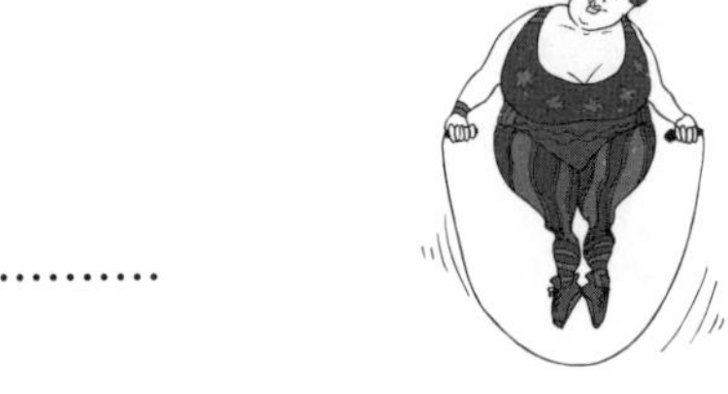

a)

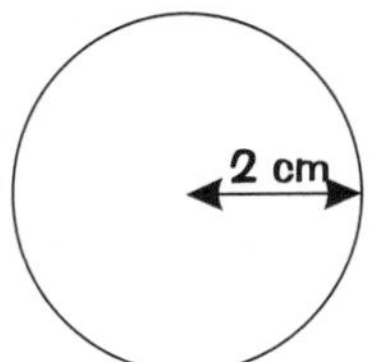

Area = $\pi \times \text{radius}^2$ = ...

b) A circle of radius 9 cm. Area = $\pi \times \text{radius}^2$ = ...

Give your answers to **c)** and **d)** to 3 sig. figs.

c)

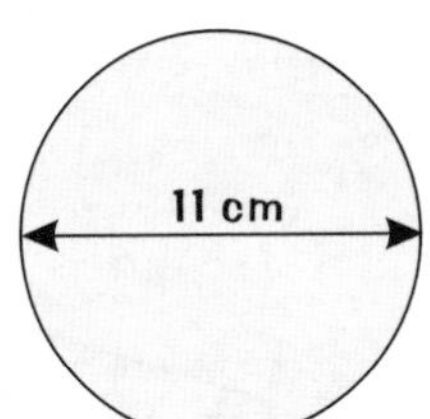

You must find the radius first.

Radius = diameter / 2 = ...

Area = ...

d) A circle of diameter 28 cm. Radius = ...

Area = ...

Q7 What is the area of this semicircular rug?
Give your answer to the nearest whole number.

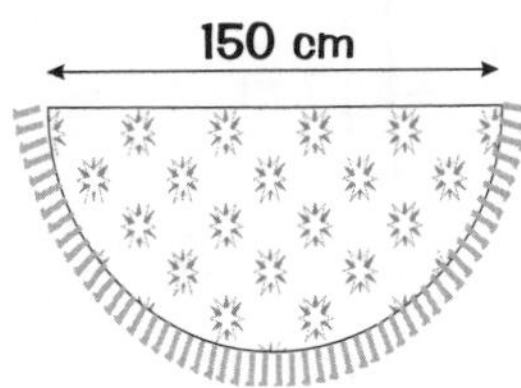

Area = ...

Q8 Patrick is using weed killer to treat his circular lawn, which has a diameter of 12 m. The weed killer instructions say to use 20 ml for every square metre of lawn to be treated. How much weed killer does he need to use? Give your answer to the nearest ml.

....

Q9 This circular pond has a circular path around it. The radius of the pond is 72 m and the path is 2 m wide. Calculate to the nearest whole number:

a) the area of the pond. ....

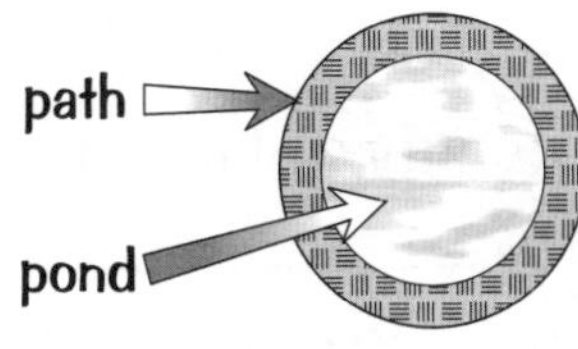

b) the area of paving stones needed to repave the path.

....

Volume

Q1 Each shape has been made from centimetre cubes. The volume of a centimetre cube is 1 cubic cm. How many cubes are there in each shape? What is the volume of each shape?

a)

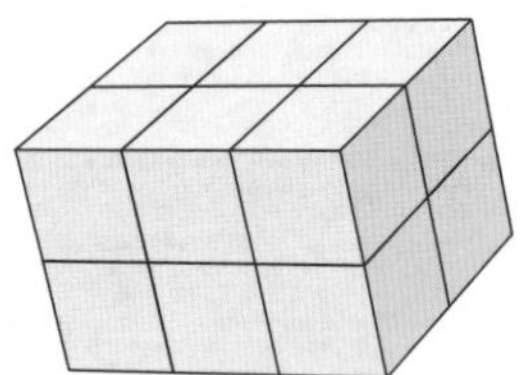

There are ...... cubes.
The volume is
...... cubic cm.

b)

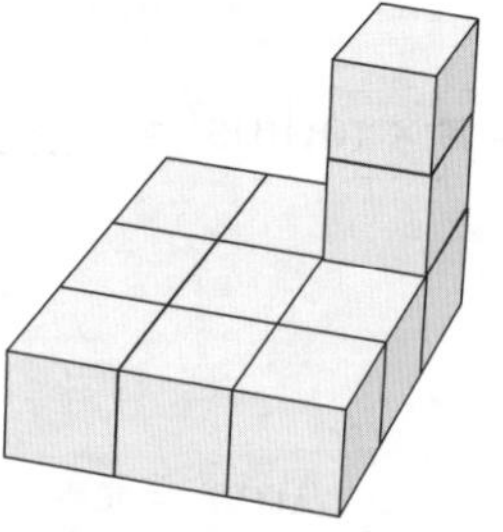

There are ...... cubes.
The volume is
...... cubic cm.

c)

There are ...... cubes.
The volume is
...... cubic cm.

d)

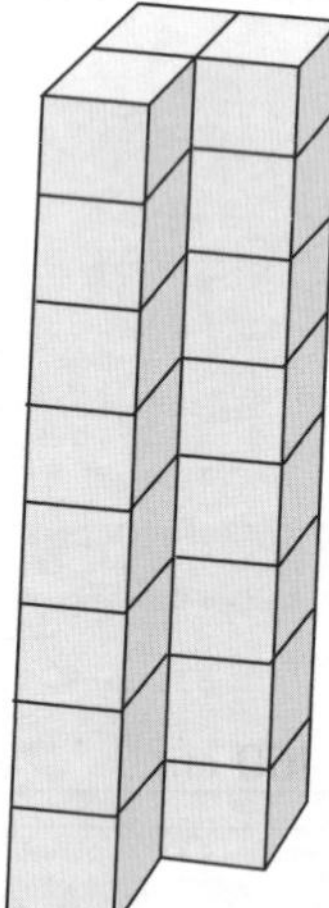

There are ...... cubes.
The volume is
...... cubic cm.

e)

There are ...... cubes.
The volume is
...... cubic cm.

f)

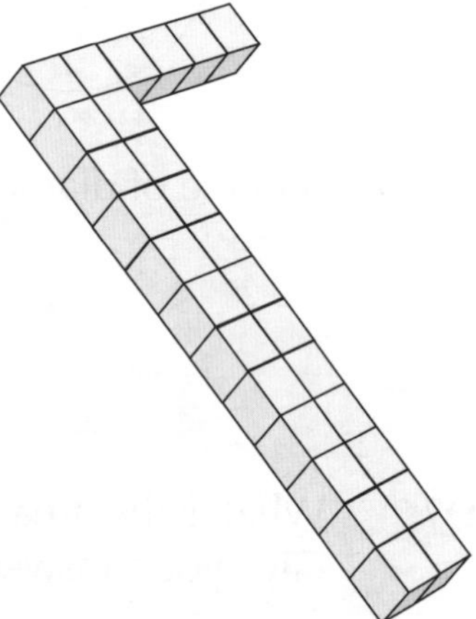

There are ...... cubes.
The volume is
...... cubic cm.

g)

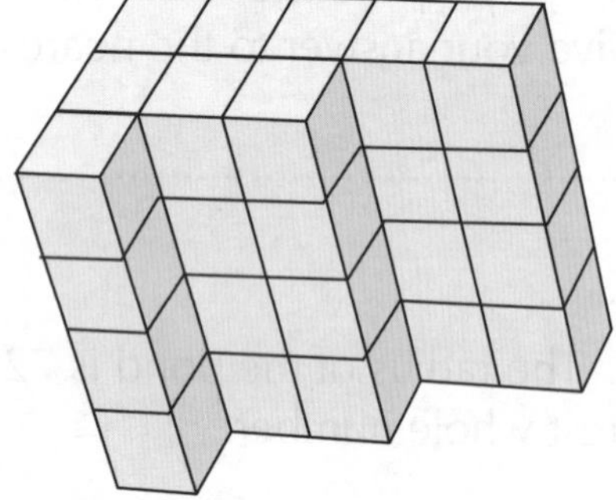

There are ...... cubes.
The volume is ...... cubic cm.

h)

There are ...... cubes.
The volume is ...... cubic cm.

You simply add up the cubes... but make sure you don't miss any — remember that there are some rows at the back too.

Volume

Finding volumes of cubes and cuboids is just like finding areas of squares and rectangles — except you've got an extra side to multiply by.

Q2 A match box measures 7 cm by 4 cm by 5 cm.
What is its volume?

....

Q3 What is the volume of a cube of side:

a) 5 cm? ....

b) 9 cm? ....

c) 15 cm? ....

Q4 Sam has a rectangular jelly mould measuring 16 cm by 16 cm by 6 cm.
Is the mould big enough to hold 1600 cm^3 of jelly?

....

Q5 A box measures 9 cm by 5 cm by 8 cm.

a) What is its volume? ....

b) What is the volume of a box twice as long, twice as wide and twice as tall? ....

Q6 Which holds more, a box measuring 12 cm by 5 cm by 8 cm or a box measuring 10 cm by 6 cm by 9 cm?

....

Q7 A rectangular swimming pool is 12 m wide and 18 m long.
How many cubic metres of water are needed to fill the pool to a depth of 2 m?

....

Q8 An ice cube measures 2 cm by 2 cm by 2 cm. What is its volume? ....
Is there enough room in a container measuring 8 cm by 12 cm by 10 cm for 100 ice cubes? ....

Volume

Contrary to popular belief, there isn't anything that complicated about prisms — they're only solids with the same shape all the way through. The only bit that sometimes takes a bit longer is finding the cross-sectional area. (Back a few pages for a reminder of areas.)

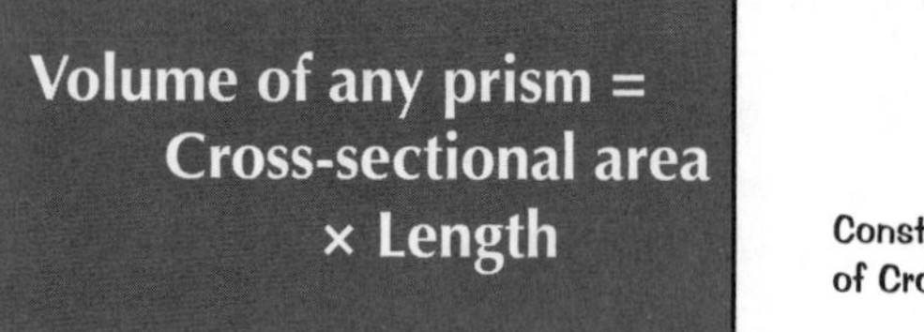

Volume of any prism =
Cross-sectional area
× Length

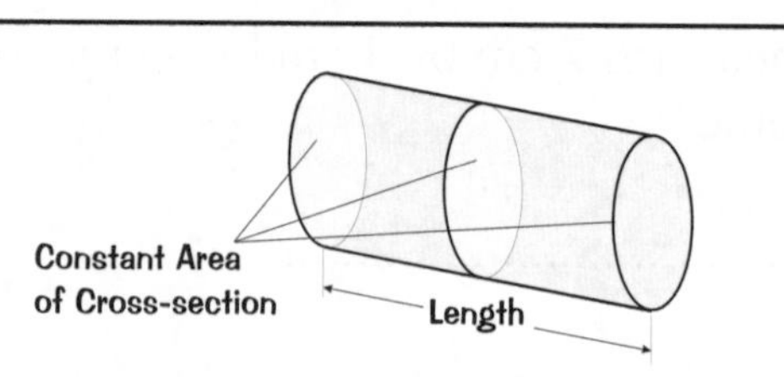

Q9 A tree trunk can be thought of as a circular prism with a height of 1.7 m. If the trunk has a diameter of 60 cm what volume of wood is this in m^3?

....

Q10 A coffee mug is a cylinder closed at one end. The internal radius is 7 cm and the internal height is 9 cm.

a) Taking π to be 3.14, find the volume of liquid the mug can hold.

....

b) If 1200 cm^3 of liquid is poured into the mug, find the depth to the nearest whole cm.

....

....

The depth is just the length of mug taken up by the liquid — which you find by rearranging the volume formula.

Q11 An unsharpened pencil can be thought of as a regular hexagonal prism with a cylinder of graphite through the centre.

You need to use a bit of Pythagoras to find the height of the triangles.

a) By considering a hexagon to be made up of six equilateral triangles, calculate the area of the cross-section of the hexagonal prism shown.

....

b) Find the area of wood in the cross-section.

....

circle 2mm diameter

hexagon 4mm each side

c) If the pencil is 20 cm long what is the volume of wood in the pencil?

....

Vertices, Faces and Edges

Q1 What are the names of these shapes?

a)

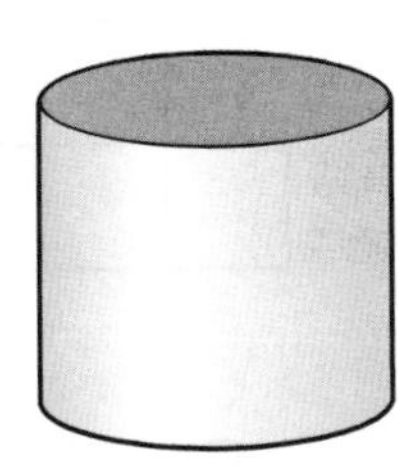

....

b)

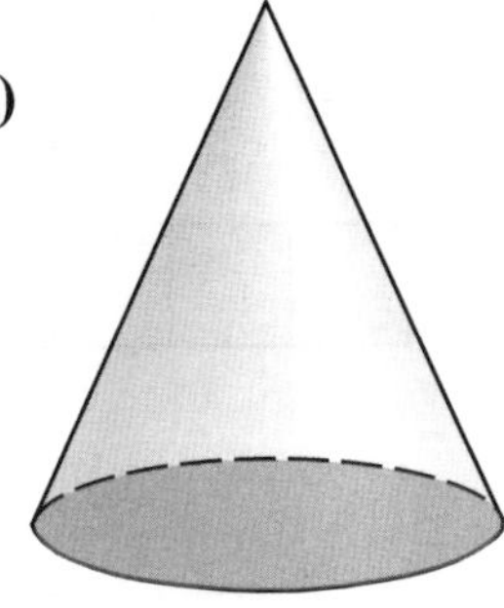

....

c)

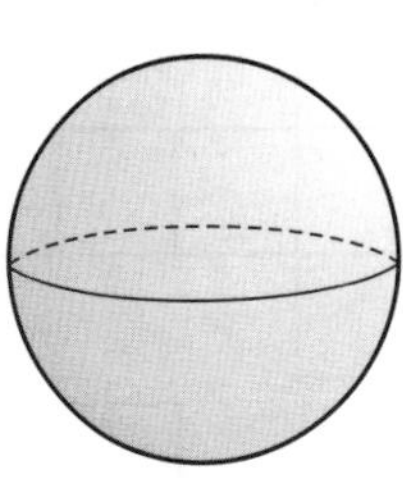

....

Q2 Fill in the boxes in the table.

Vertex just means corner.

	Name of SHAPE	Number of FACES	Number of EDGES	Number of VERTICES

Name that shape... they're really keen on putting these in the Exam — gonna have to get learning those shape names, aren't you...

Solids and Nets

Q1 Which of the following nets would make a cube? ....

a)

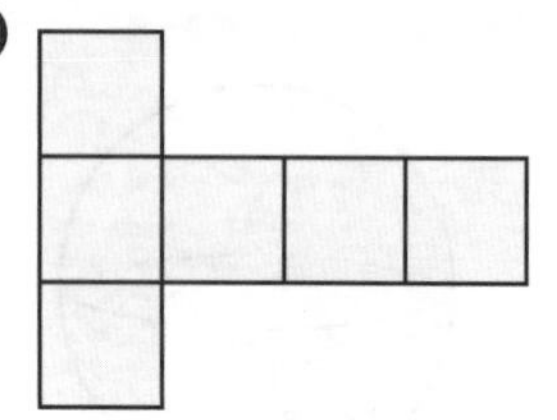

b)

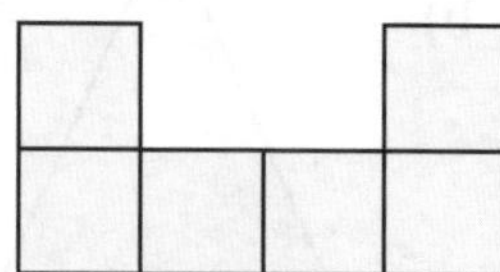

c)

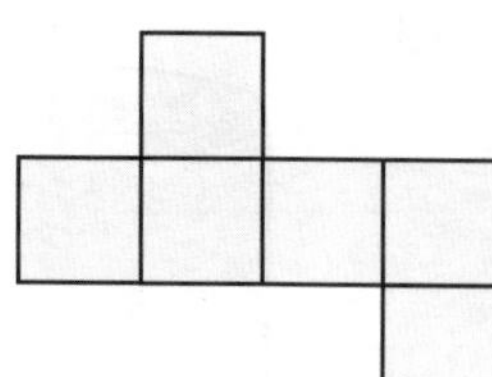

d)

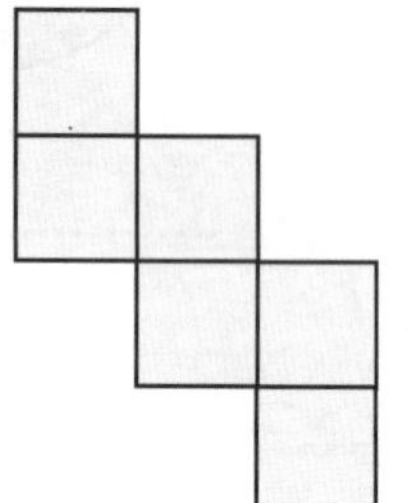

e)

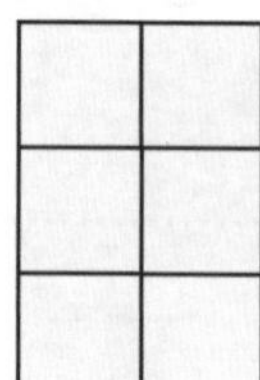

f)

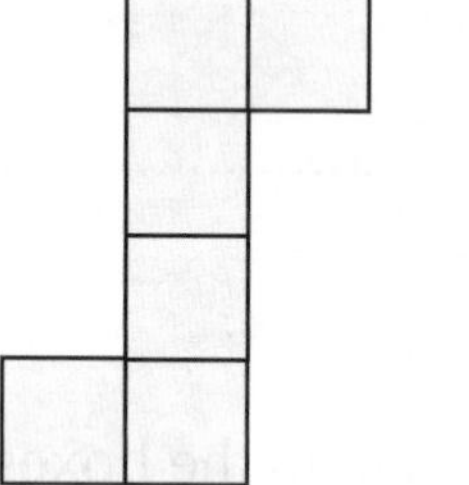

Q2 On a separate piece of paper, draw an accurate net for each of the following solids.

a)

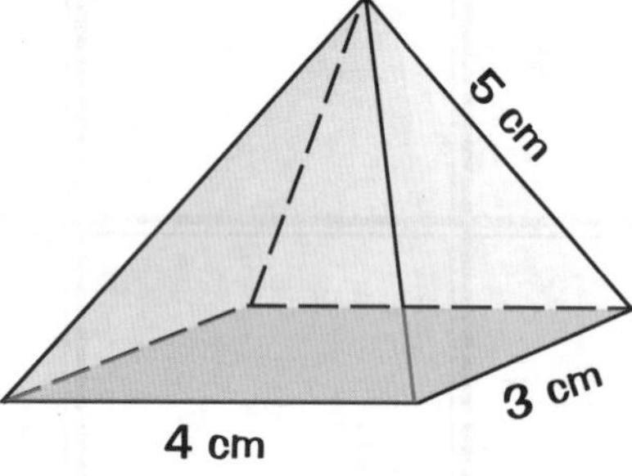

b)

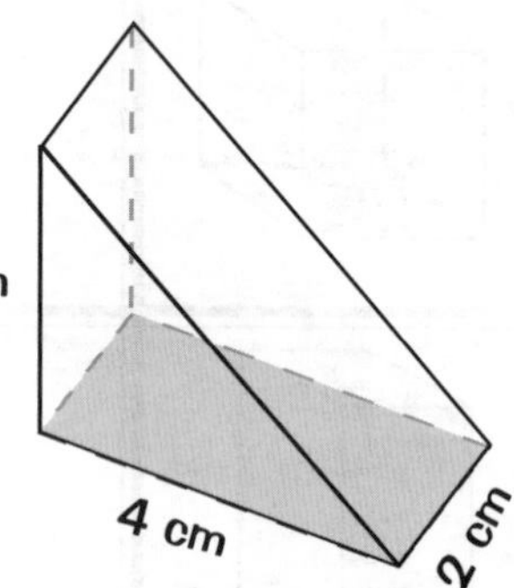

Q3 This net will make a common, mathematical solid.
Name the solid:

....

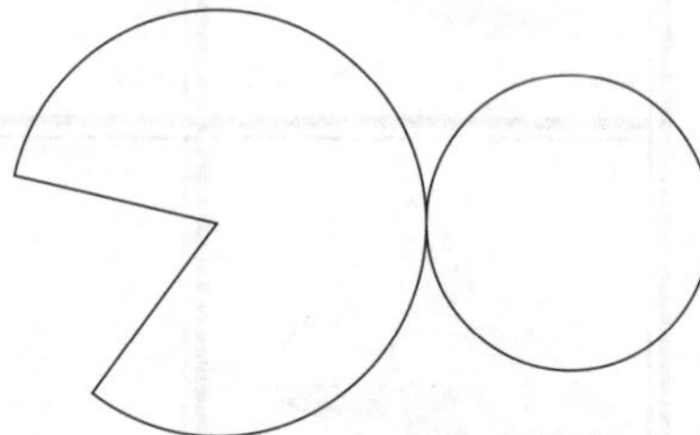

You've got to think about folding the net up to make the shape — if you're struggling, the best thing to do is practise your origami skills...

Angle Rules

Hope you've learnt those angle rules for a straight line and round a point...

Q1 Work out the angles labelled:

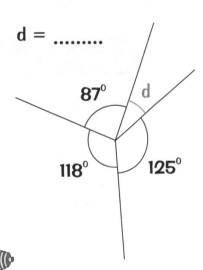

a = ...

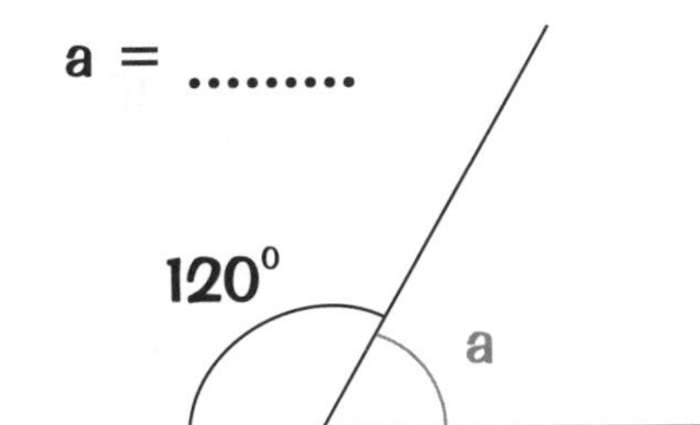

b = ...

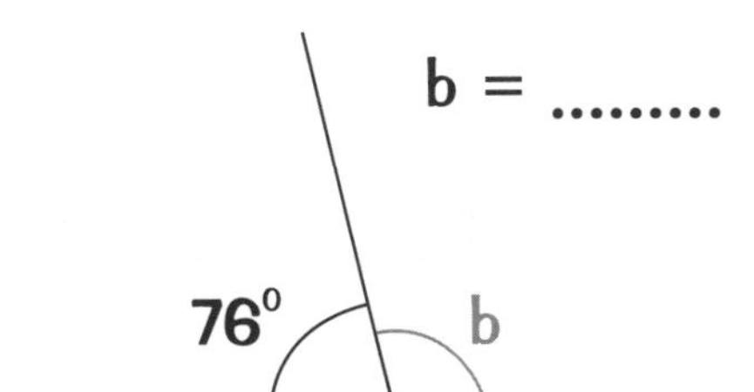

c = ...

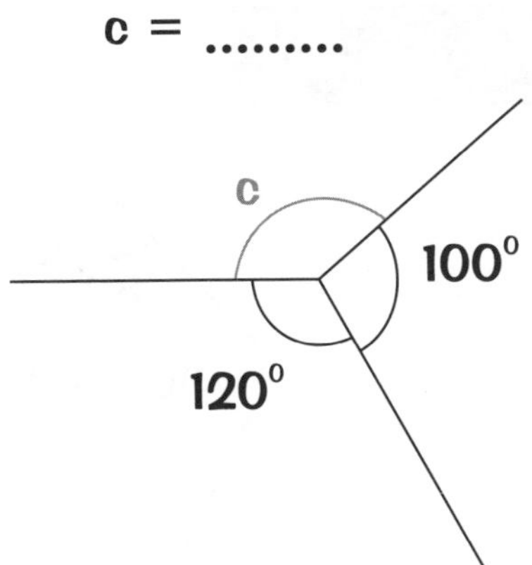

d = ...

87°
d
118°
125°

e = ...

f = ...

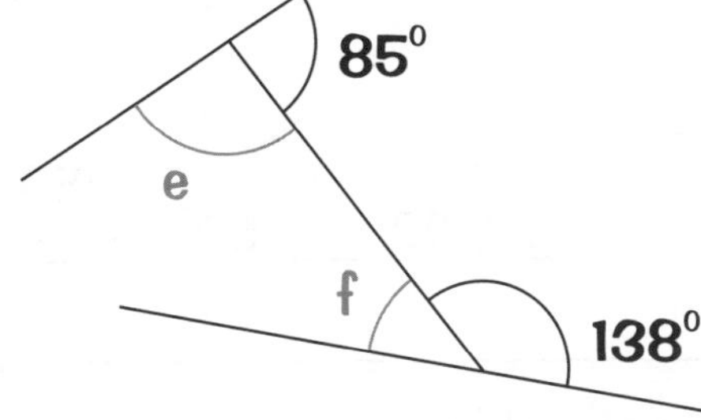

g = ...

h = ...

i = ...

g
41°
h
53°
i

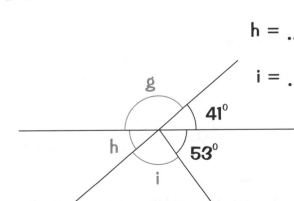

Angle Rules

The three angles inside a triangle always add up to 180°

Q2 Work out the missing angle in each of these triangles. The angles are not drawn to scale so you cannot measure them.

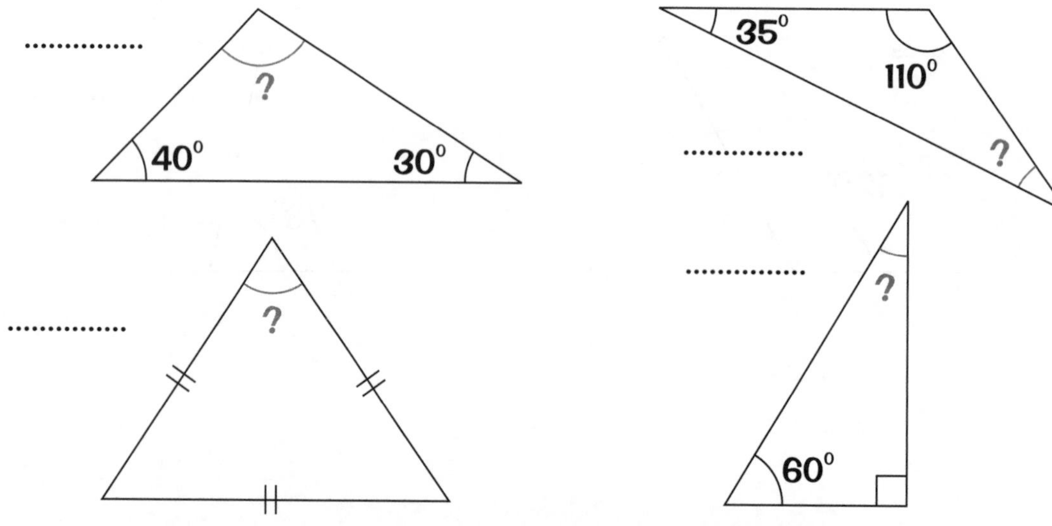

The angles in a quadrilateral always add up to 360°

Q3 Work out the missing angles in these quadrilaterals.

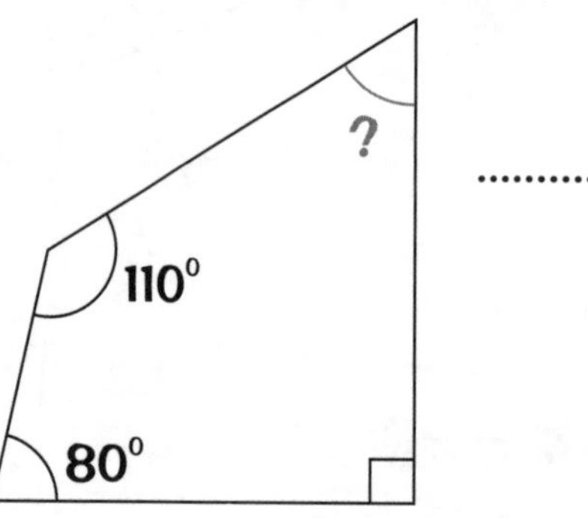

....

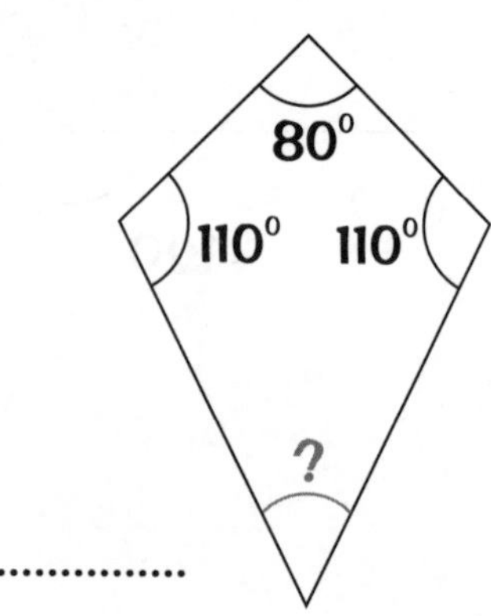

....

Q4 Work out the missing angles in these diagrams.

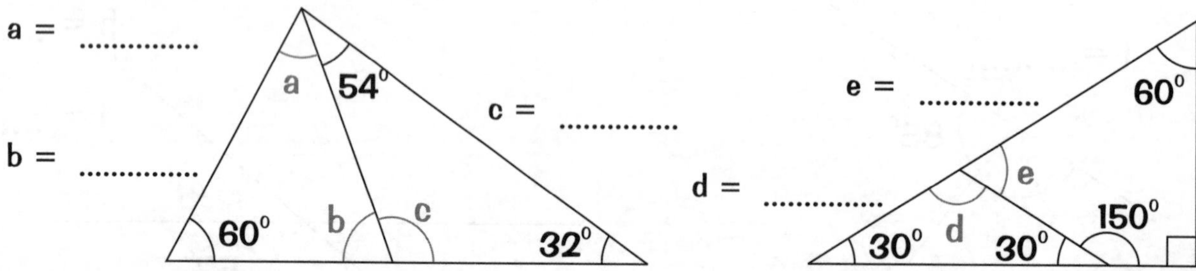

You'd better get learning these rules too — they're not that hard, and you'll be well and truly stumped without them.

Parallel Lines

Once you know the 3 ANGLE RULES for parallel lines, you can find all the angles out from just one — ah, such fun...

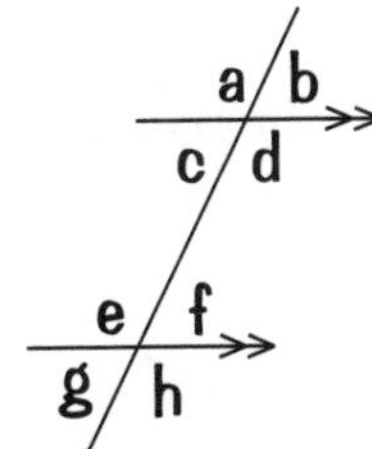

c = f and d = e — Alternate angles

a = e, c = g, b = f and d = h — Corresponding angles

d + f = 180⁰, c + e = 180⁰ — Supplementary angles

Find the sizes of the angles marked by letters in these diagrams. For each one, write down whether it is part of a pair of alternate, corresponding or supplementary angles.

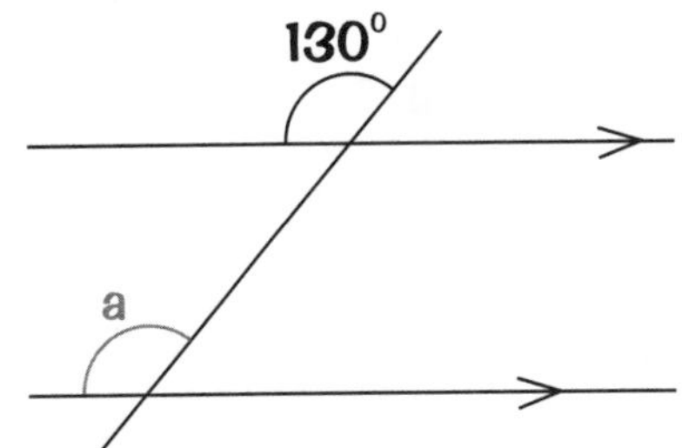

a = ...

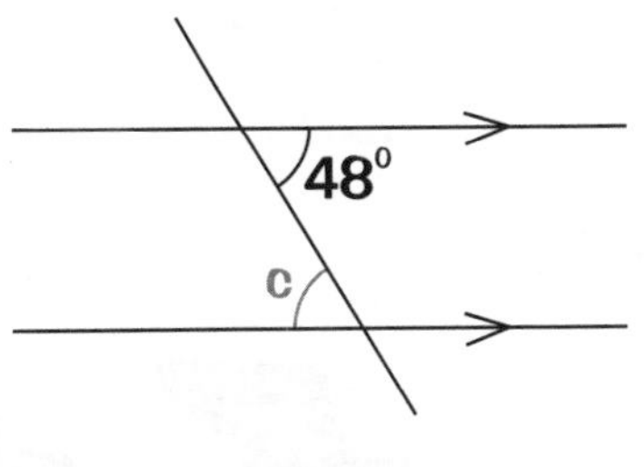

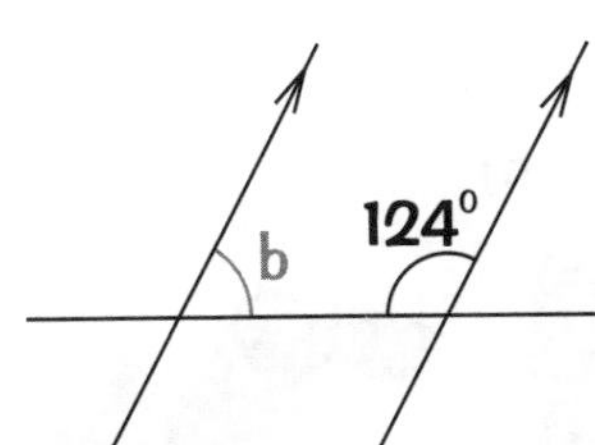

b = ...

c = ...

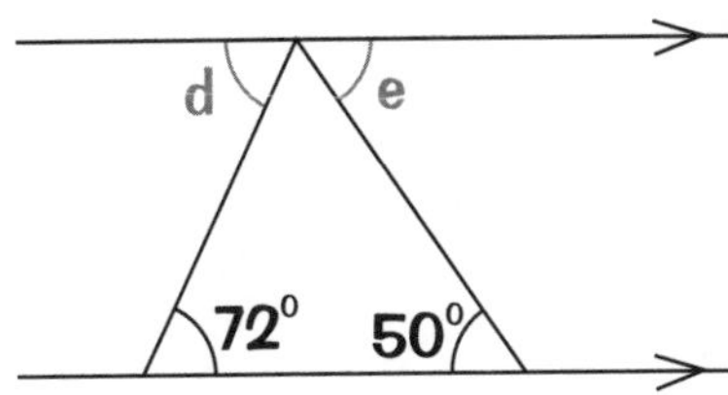

d = ...

e = ...

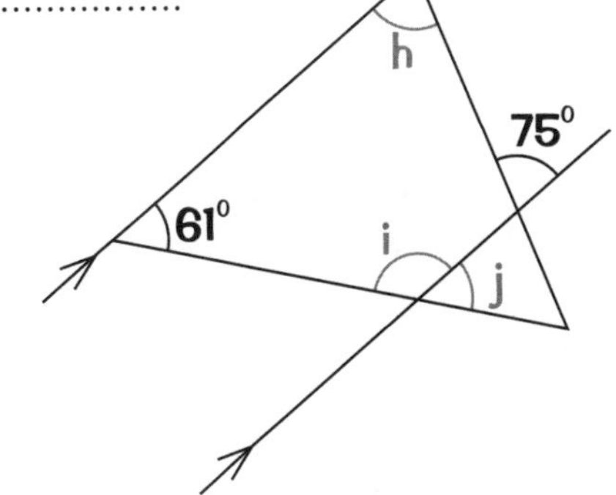

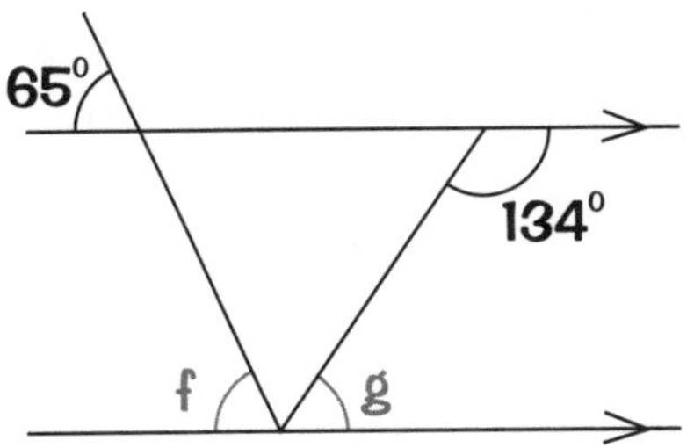

h = ...

f = ...

i = ...

g = ...

j = ...

Congruence

I reckon these are pretty easy — so while you're racing through them, you can be thinking: "3 shapes are CONGRUENT (<u>exactly the same</u>) and 1 isn't."

In each of the following sets of shapes underline the one which is not congruent.

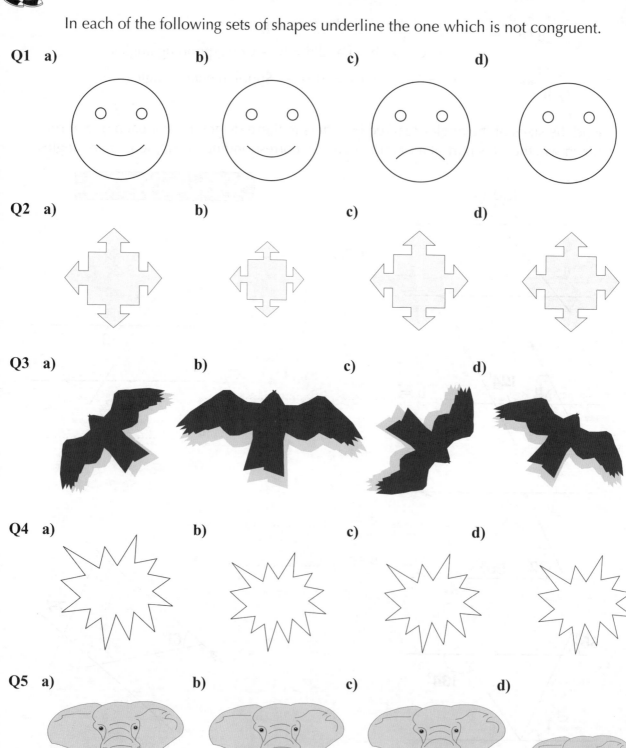

Q1 **a)**  **b)**  **c)**  **d)** 

Q2 **a)** 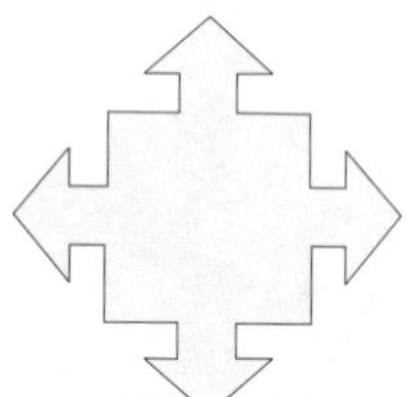 **b)** 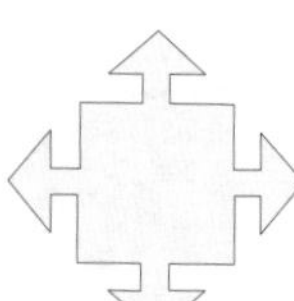 **c)** 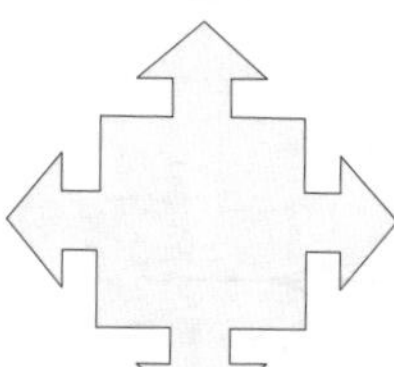 **d)** 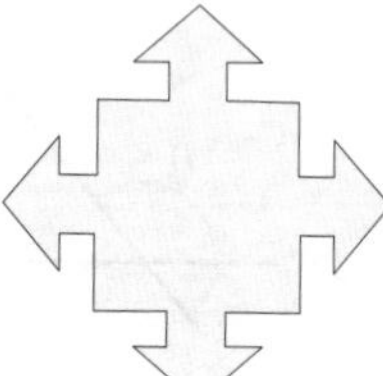

Q3 **a)**  **b)** 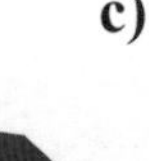 **c)** **d)** 

Q4 **a)**  **b)**  **c)**  **d)** 

Q5 **a)**  **b)**  **c)**  **d)** 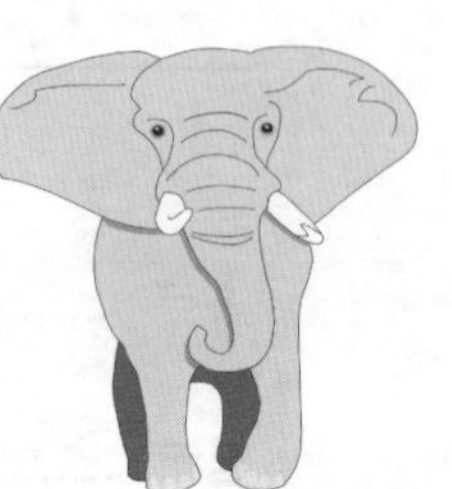

Similarity and Enlargement

Similarity and Scale Factor

Two shapes are <u>similar</u> if they're the <u>same shape</u> but different size. The lengths of the two shapes are related to the scale factor by this very important formula triangle:

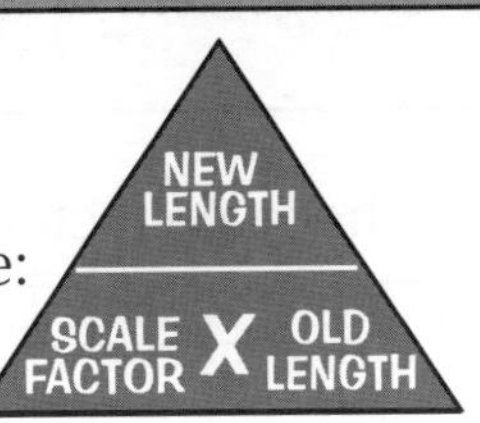

Q1 Two picture frames are shown. One picture is <u>similar</u> to the other. Calculate L cm, the length of the smaller frame.

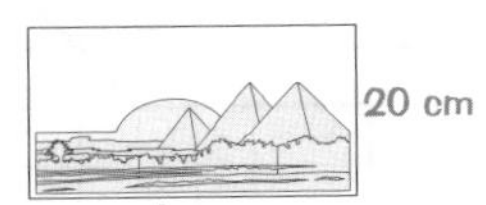

....

Q2 For each of the following pairs, say with a reason whether the shapes are similar.

i)

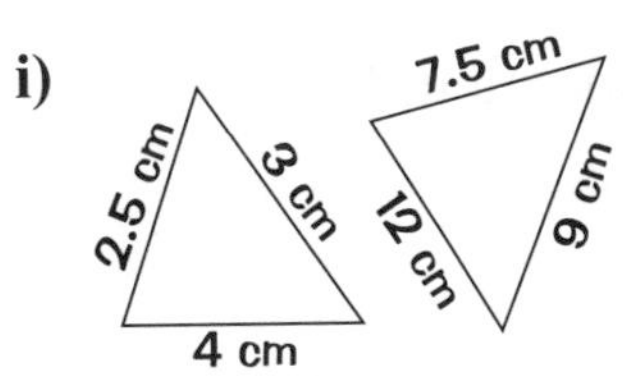

....

ii)

2 cm
48° 78°
5.5 cm
9 cm
78° 48°
3 cm

....

iii)

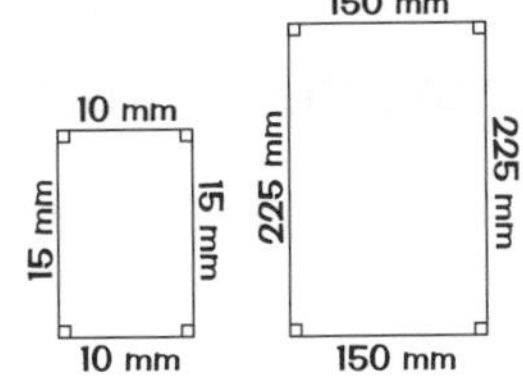

....

iv)

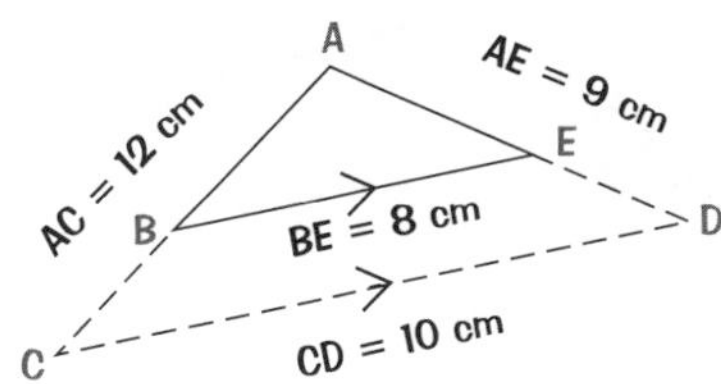

....

Q3 Angle ABC = Angle PQR and Angle BCA = Angle QRP

a) Are the triangles similar? ....

Calculate the following lengths:

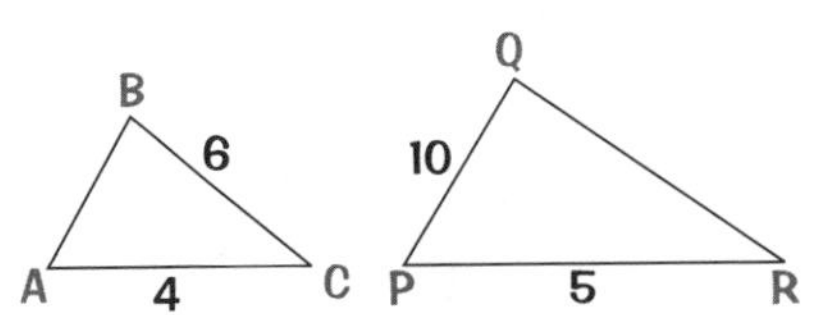

b) AB ....   **c)** QR ....

Q4 Which of the following must be <u>similar</u> to each other? ....

A Two circles
B Two rhombuses
C Two rectangles
D Two squares
E Two equilateral triangles
F Two isosceles triangles

Transformations — Enlargements

The scale factor (see previous page) is a fancy way of saying HOW MUCH BIGGER the enlargement is than the original.

Q1 What is the scale factor of each of these enlargements?

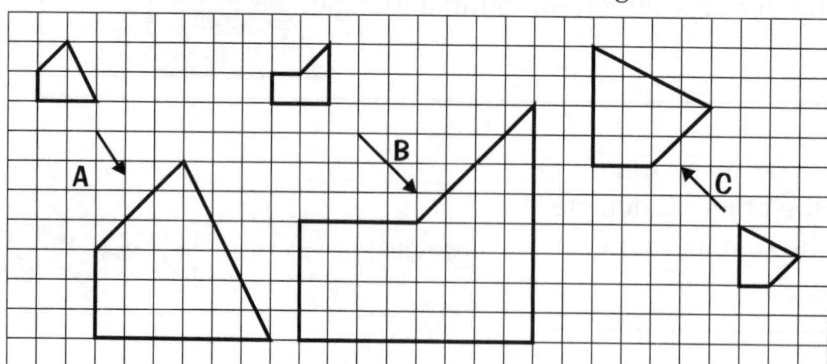

Just pick one of the sides and see how many times longer it is.

A: Scale factor is .... **B:** Scale factor is .... **C:** Scale factor is ....

Q2 Enlarge this triangle using scale factor 4 and centre of enlargement C.

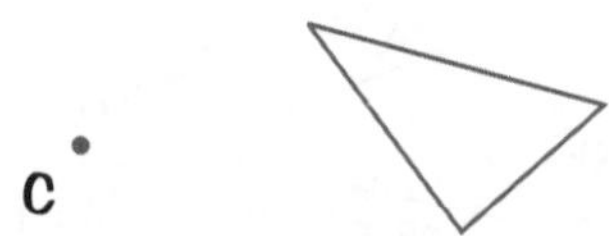

Q3 **a)** Enlarge this shape using scale factor 3 and centre of enlargement E.

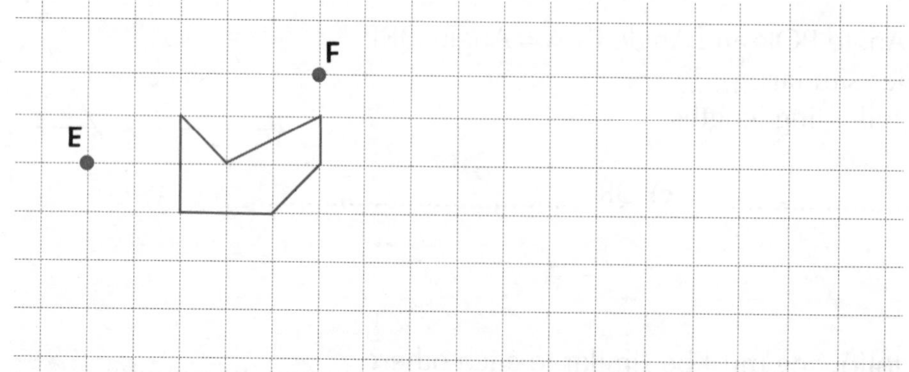

b) Now enlarge the original shape using scale factor 2 and centre of enlargement F.

Transformations — Translation

Translations can be described using vectors.

The vector $\binom{2}{5}$ means move 2 spaces to the right and 5 spaces up.

The vector $\binom{-3}{-4}$ means move 3 spaces to the left and 4 spaces down.

Q1 Translate the shapes A, B and C using these vectors: A $\binom{-4}{-3}$ B $\binom{5}{5}$ C $\binom{4}{-4}$
Label the images A′, B′ and C′.

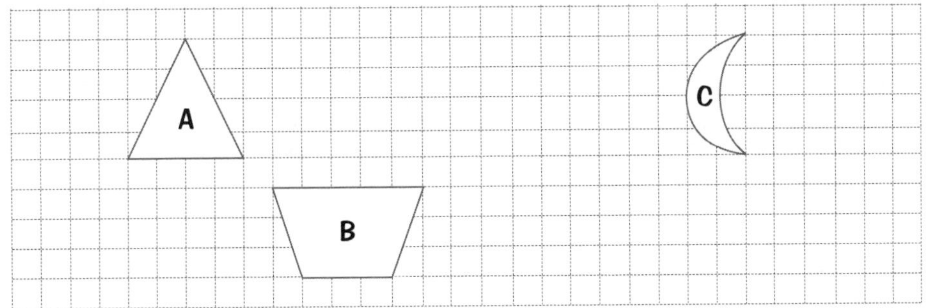

Q2 Translate shape A below using the vectors given in order, drawing the image each time:

P $\binom{3}{4}$ Q $\binom{9}{2}$

R $\binom{3}{-4}$ S $\binom{-8}{-4}$

Label the images A′, A″, A‴, A⁗.

Q3 Write down the translation vectors for the translations shown.

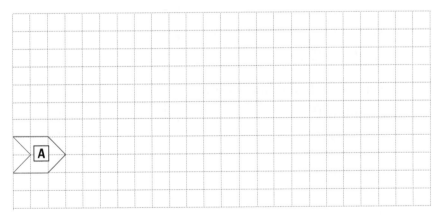

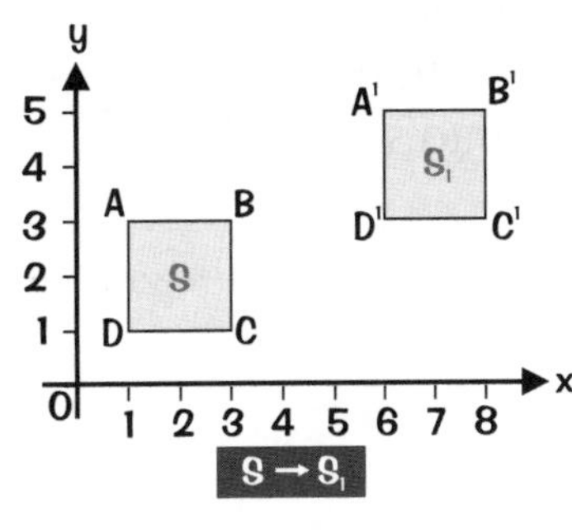

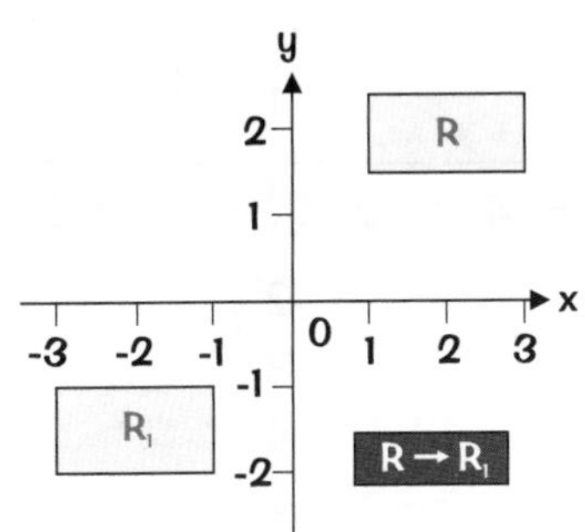

a) .... b) .... c) ....

Transformations — Reflection

Q1 Reflect each shape in the line $x = 4$.

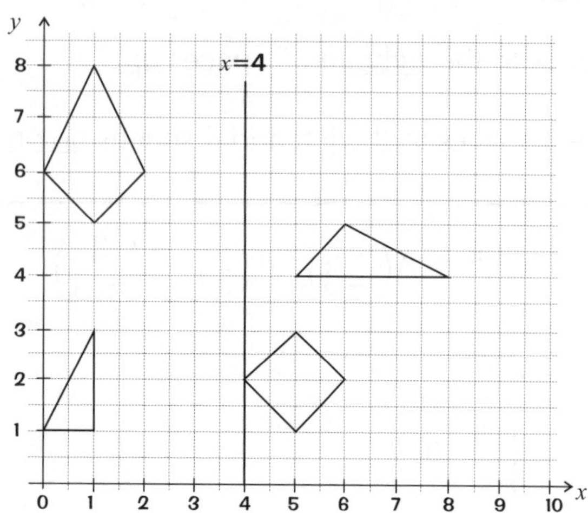

Q2 Reflect the shapes in the line $y = x$.

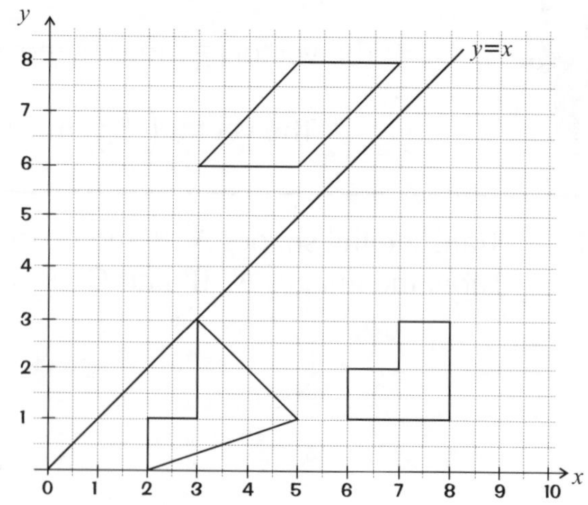

Q3 Reflect ① in the line $y = 5$, label this ②.

Reflect ② in the line $x = 9$, label this ③.

Reflect ③ in the line $y = x$, label this ④.

Reflect ④ in the line $x = 4$, label this ⑤.

Reflect ⑤ in the line $y = x$, label this ⑥.

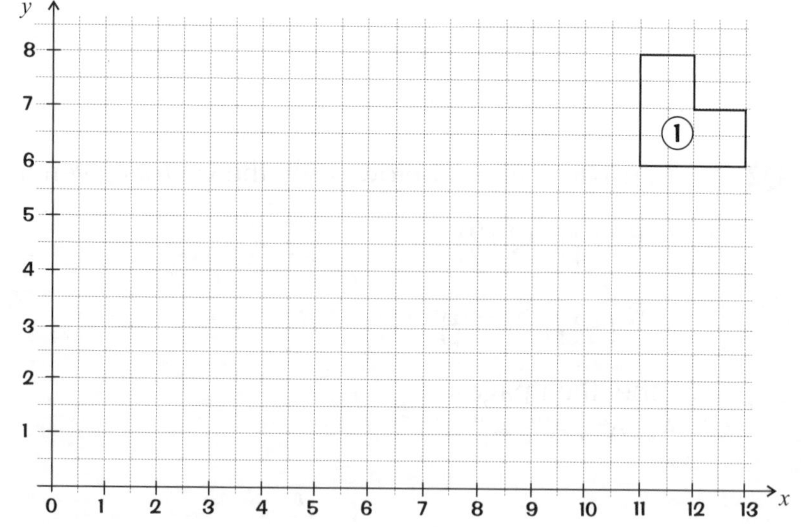

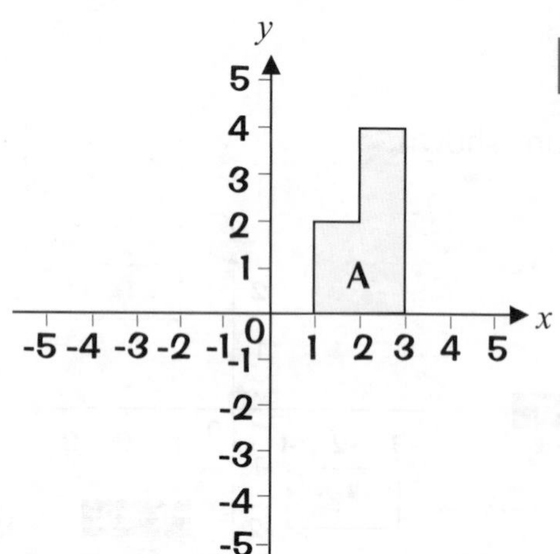

Q4 **a)** Draw the result of reflecting shape A in the x-axis, label this A′.

b) Draw the result of reflecting shape A′ in the y-axis, label this A″.

c) What single transformation would turn shape A into shape A″?

....

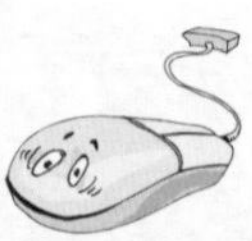

Nothing fancy here, is there? Reflection's just mirror drawing really. And we've all done that before...

Transformations — Rotation

Q1 The centre of rotation for each of these diagrams is ***X***. Rotate (turn) each shape as asked then draw the new position of the shape onto each of the diagrams below.

a) 180° (or ½ turn).

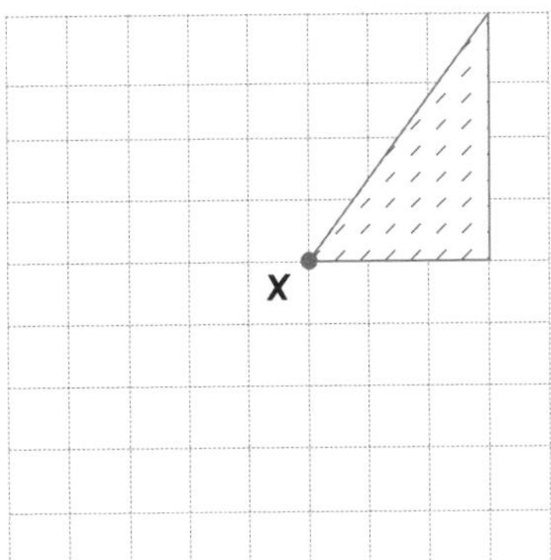

b) 270° anticlockwise (or ¾ turn anticlockwise).

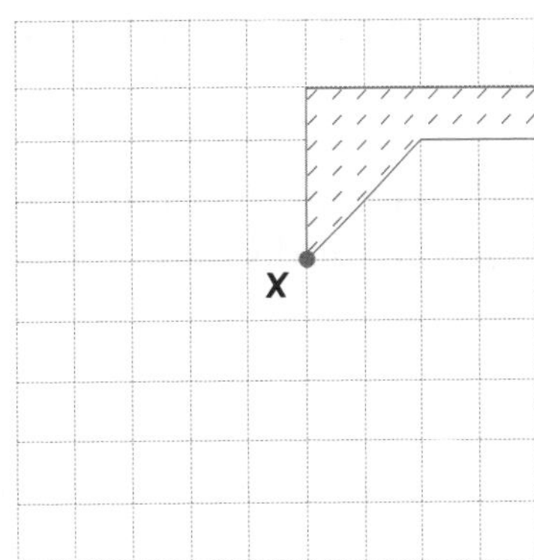

Q2 This is a scalene triangle PQR. The centre of rotation is the origin 0.

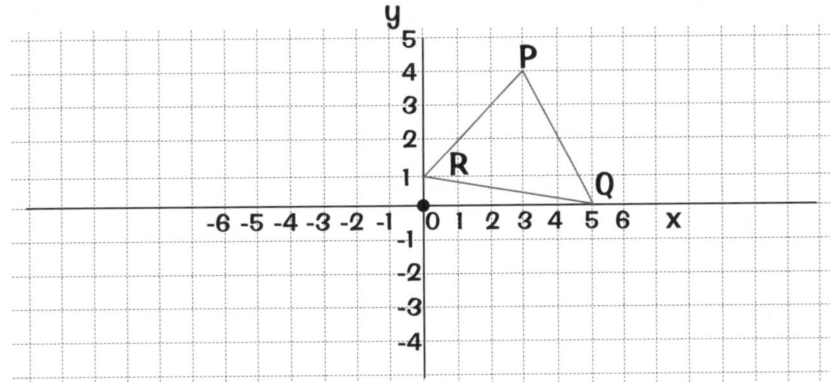

a) Write down the coordinates of...

P .... Q .... R ....

b) Rotate the triangle 90° anticlockwise about the origin 0.
Label the new triangle P′ Q′ R′.

c) Write down the coordinates of ...

P′ .... Q′ .... R′ ....

A ½ turn clockwise is the same as a ½ turn anti-clockwise — and a ¼ turn clockwise is the same as a ¾ turn anti-clockwise. Great fun, innit...

Pythagoras' Theorem

If you're as big a fan of Pythagoras as me, you'll ignore him and use this method instead:

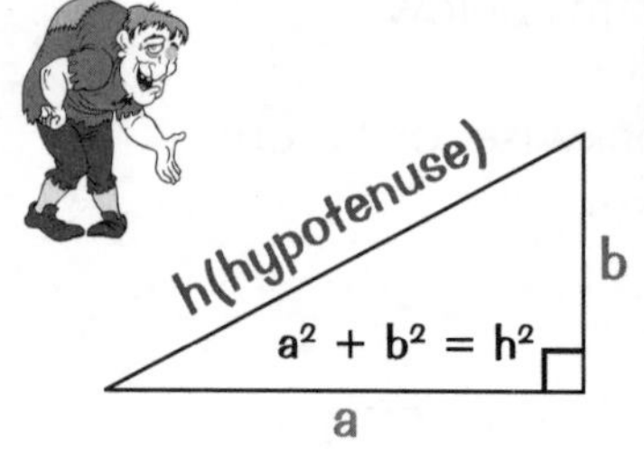

The Simple Three Step Method

1) SQUARE the two numbers that you are given.
2) To find the longest side, ADD the two squared numbers.
   To find a shorter side, SUBTRACT the smaller one from the larger.
3) Take the SQUARE ROOT. Then check that your answer is sensible.

Q1 Using Pythagoras' theorem, calculate the length of the third side in these triangles.

c^2 = ... + ... = ..., c = ...

d^2 = ... − .... = ..., d = ...

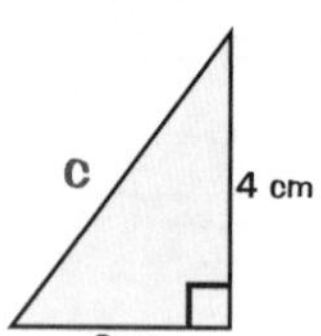

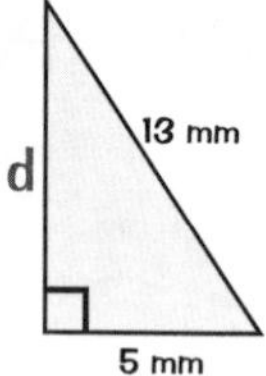

Q2 Using Pythagoras' theorem, work out which of these triangles have right-angles.

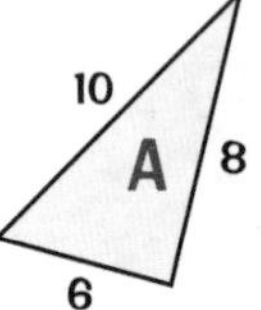

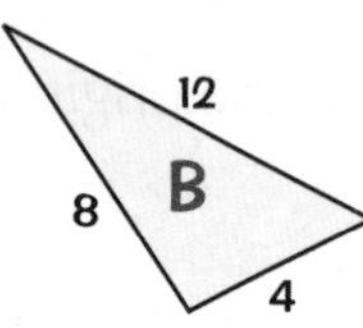

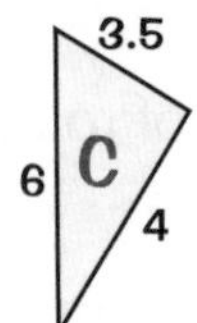

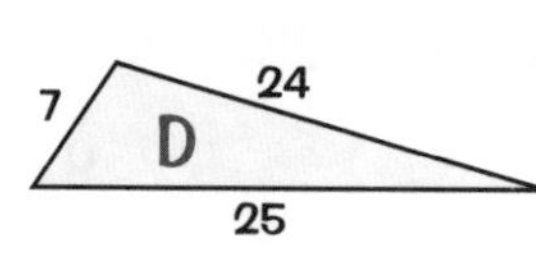

If they don't have right-angles, the numbers won't fit the formula.

....

....

Q3 Calculate the missing lengths in these quadrilaterals. Give your answers to 3 sig. figs.

e = ...

f = ...

g = ...

h = ...

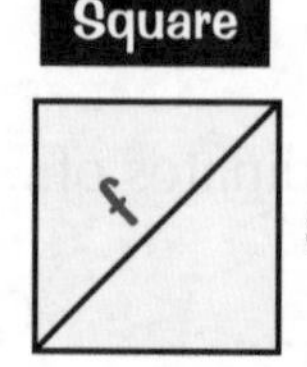

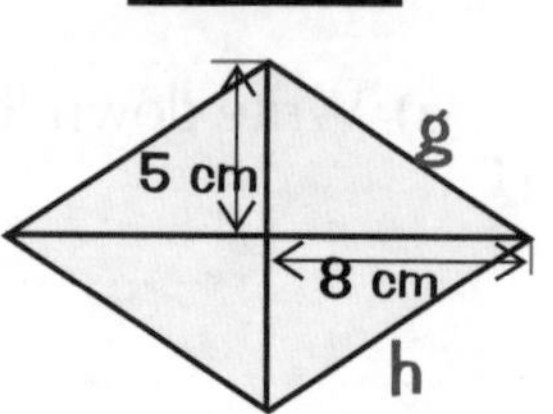

Q4 A window cleaner needs to clean the upstairs windows of an office. He has a ladder 10 m long, but for safety reasons he can only put the bottom of it a minimum of 3 m away from the wall. In whole metres, what is the maximum height that the top of the ladder can reach when used safely?

....

....

....

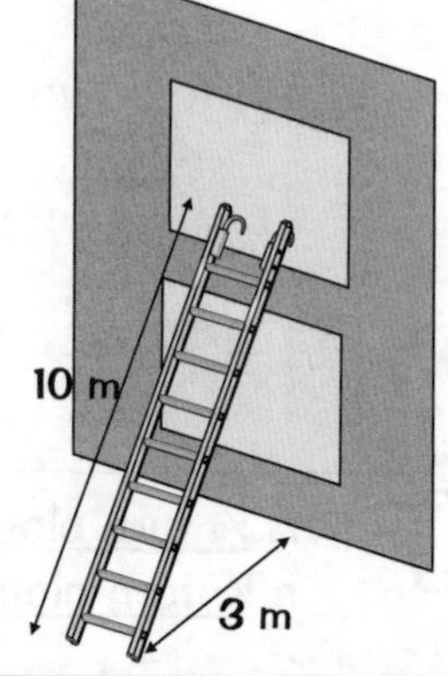

Formula Triangles

You can use a formula triangle for ANY FORMULA with THREE THINGS, where two are MULTIPLIED to give the third.

Eg: area of a rectangle = length × height.
This will give the formula triangle:

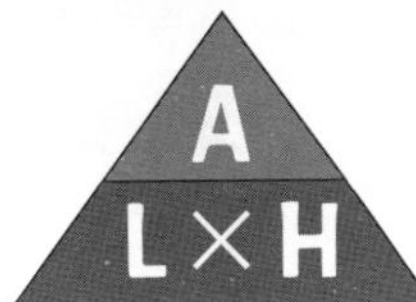

... All you do is cover up what you want with your finger and the other two bits will tell you how to calculate it. Couldn't be simpler....

Q1 The formula for finding the area of a triangle is Area = half × base × height, i.e.

$$A = \left(\frac{b}{2}\right) \times h$$

Draw a formula triangle and use it to find:

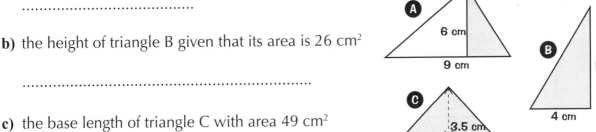

a) the area of triangle A

....

b) the height of triangle B given that its area is 26 cm^2

....

c) the base length of triangle C with area 49 cm^2

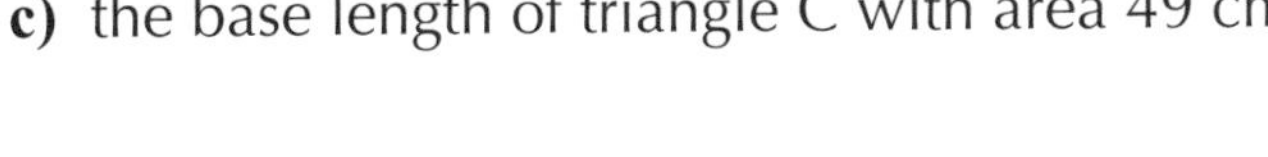

....

Q2 The circumference of a circle is given by the formula $c = \pi d$, where c = circumference, d = diameter. Draw a formula triangle and use it to find:

a) the circumference of a bike wheel with a diameter of 72 cm ....

b) the diameter of a jam jar with a circumference of 21 cm ....

Q3 A formula used by accountants is $L = S/Q$. L is Lateral Forecast, S is the Spend Parameter and Q is the Quotient Charter. Draw a formula triangle relating L, S and Q and use it to find:

a) the Lateral Forecast when the Spend Parameter is 120 and the Quotient Charter is 8 ....

b) the Quotient Charter when the Spend Parameter is 408 whilst the Lateral Forecast is 24 ....

Don't worry about what the funny words mean — it's just a formula triangle question.

Speed

This is an easy enough formula — and of course you can put it in that good old formula triangle as well.

$$\text{Average speed} = \frac{\text{Total distance}}{\text{Total time}}$$

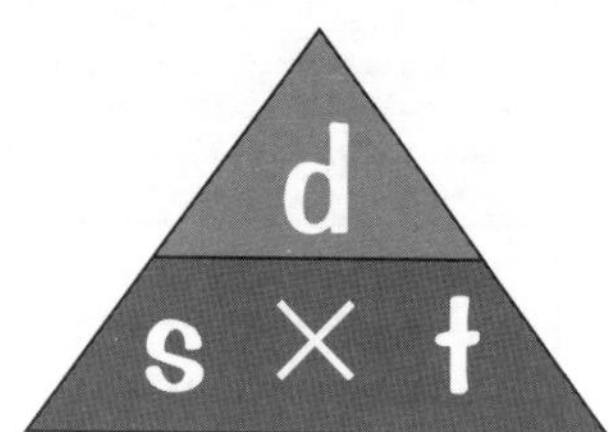

Q1 A motorbike travels for 3 hours at an average speed of 55 mph. How far has it travelled?

....

Q2 Complete this table.

Distance Travelled	Time taken	Average Speed
210 km	3 hrs	
135 miles		30 mph
	2 hrs 30 mins	42 km/h
9 miles	45 mins	
640 km		800 km/h
	1 hr 10 mins	60 mph

Q3 An athlete can run 100 m in 11 seconds.
Calculate the athlete's speed in:

a) m/s

....

b) km/h

....

Q4 Simon is driving on the motorway at an average speed of 67 mph.
He sees a sign telling him that he is 36 miles away from the next service station.

a) To the nearest minute, how long will it take Simon to reach the service station?

....

b) Later, Simon drove through some roadworks where the speed limit was 50 mph. Two cameras recorded the time taken to travel 1200 m through the roadworks as 56 seconds. Was Simon speeding through the roadworks?

1 mile = 1.6 km

....

Scale Drawings

Q1 The scale on this map is 1 cm : 4 km.

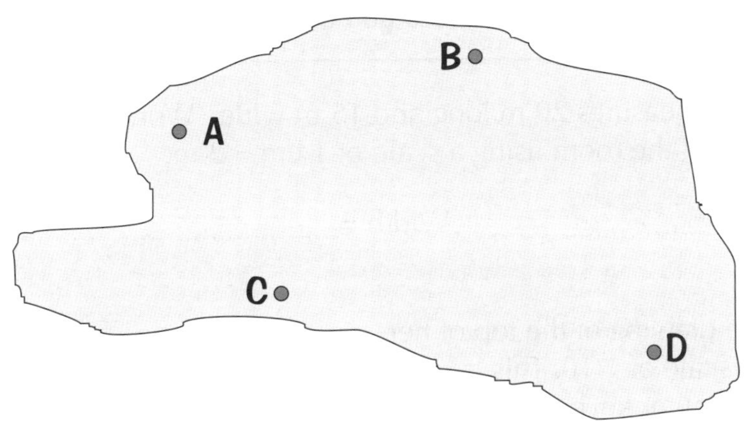

a) Measure the distance from A to B in cm. ....

b) What is the actual distance from A to B in km? ....

c) A helicopter flies on a direct route from A to B, B to C and C to D.
What is the total distance flown in km?

....

Q2 Frank has made a scale drawing of his garden to help him plan some improvements.
The scale on the drawing is 1 : 50.

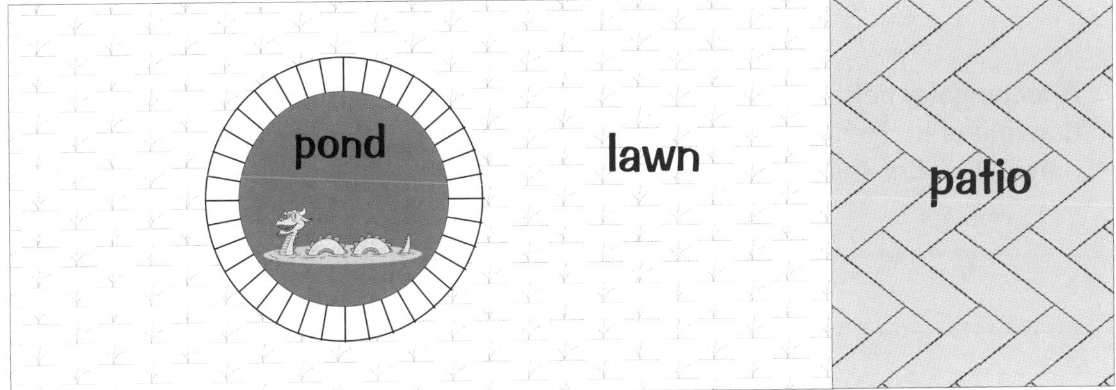

a) Frank wants to put up a fence along the three outside edges of the lawn.
How many metres of fencing does he need to buy?

....

b) What are the actual dimensions of Frank's patio in m?

....

If the scale doesn't say what units it's in, it just means that both sides of the ratio are the same units — so 1 : 1000 would mean 1 cm : 1000 cm.

Scale Drawings

Watch out for those units... there's quite a mixture here — you'll have to convert some of them before you can go anywhere.

Q3 A rectangular room measures 20 m long and 15 m wide. Work out the measurements for a scale drawing of the room using a scale of 1 cm = 2 m.

Length = ... Width = ...

Q4 Katie drew a scale drawing of the top of her desk. She used a scale of 1:10. This is her drawing of the computer keyboard. What are the actual dimensions of it?

Length = ... Width = ...

Q5 Cupboards | Oven | Cupboards

This is a scale drawing of part of Paul's kitchen. Measure the width of the gap for the oven.

.... mm.

The drawing uses a scale of 1 : 60.
Work out the maximum width of oven, in cm, that Paul can buy.

....

Q6 A rectangular field is 60 m long and 40 m wide. The farmer needs to make a scale drawing of it. He uses a scale of 1 : 2000. Work out the measurements for the scale drawing. (Hint — change the m to cm.)

....

....

Q7 A rectangular room is 4.8 m long and 3.6 m wide. Make a scale drawing of it using a scale of 1 cm to 120 cm. First work out the measurements for the scale drawing.

Length = ...

Width = ...

On your scale drawing mark a window, whose actual length is 2.4 m, on one long wall and mark a door, actual width 90 cm, on one short wall.

Window = ...

Door = ...

Conversion Factors

The method for these questions is very easy so you might as well learn it...

1) Find the Conversion Factor
2) Multiply by it AND divide by it
3) Choose the common sense answer

The conversion factor is the link between the two units — e.g. there are 100 cm in a m so the conversion factor is 100.

Q1 Fill in the gaps using the conversion factors:

20 mm = ... cm   82 mm = ... cm   .... mm = 6 cm

9000 m = ... km   3470 m = ... km   .... m = 2 km

3 km = ... cm   .... mm = 3.4 m   .... cm = 0.5 km

6200 mg = ... g   8550 g = ... kg   2.3 kg = ... g

12 000 000 mg = ... kg   1.2 l = ... ml   4400 ml = ... l

Q2 Justin is shopping online. He looks up the following exchange rates:

1.60 US Dollars ($) to £1 Sterling.
150 Japanese Yen (¥) to £1 Sterling.
2 Australian Dollars (AUD) to £1 Sterling.

Use these exchange rates to calculate to the nearest penny the cost in Sterling of each of Justin's purchases:

a) A book costing $7.50

b) An MP3 player costing ¥7660

c) An electric guitar costing 683 AUD

....

Justin has two quotes for the cost of shipping his guitar from Australia to the UK:
155 AUD from an Australian courier and £76.45 from a British courier.

d) Which company is cheaper?

....

Q3 1 pint = 0.5714 litres. Which is better value,
2 pints of orange juice for £1.20 or 1 litre of orange juice for 95p?

Start by changing the 2 pints to litres.

....

Metric and Imperial Units

You'd better learn ALL these conversions — you'll be well and truly scuppered without them.

APPROXIMATE CONVERSIONS

1 kg = 2.2 lbs	1 gallon = 4.5 l
1 inch = 2.5 cm	5 miles = 8 km

Q1 The table shows the distances in miles between 4 cities in Scotland. Fill in the blank table with the equivalent distances in kilometres.

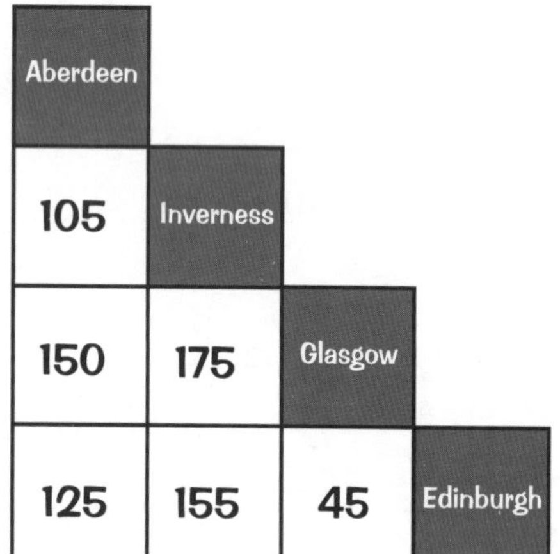

Aberdeen			
105	Inverness		
150	175	Glasgow	
125	155	45	Edinburgh

Aberdeen			
	Inverness		
		Glasgow	
			Edinburgh

Q2 Change each of these weights from kilograms to pounds.

10 kg = ... lbs   16 kg = ... lbs   75 kg = ... lbs

Change each of these capacities in gallons to litres.

5 galls = ... l   14 galls = ... l   40 galls = ... l

Q3 Convert the measurements of the note book and pencil to centimetres.

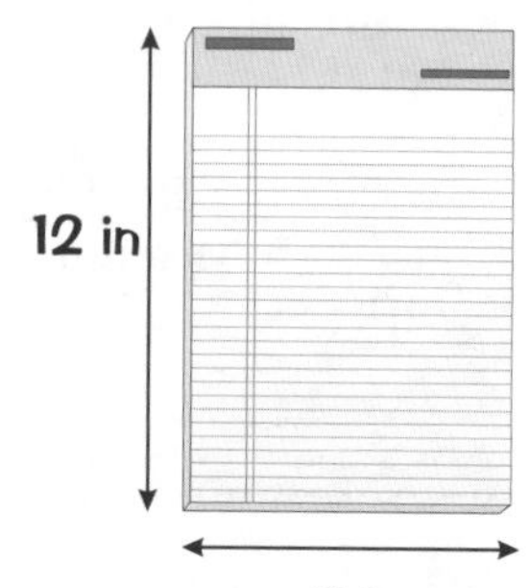

12 in = ... cm

8 in = ... cm

5 in = ... cm

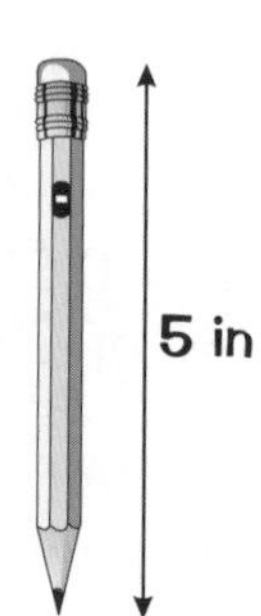

Metric and Imperial Units

Q4 The water butt in my garden holds 20 gallons of rain-water. How many litres is this?

....

Q5 Tom walked 17 km in one day, while Paul walked 10 miles. Who walked further?

....

....

It doesn't matter which distance you convert — but here it's easier to convert Paul's miles to km.

Q6 A recipe for strawberry jam requires 4 lb of sugar. How many 1 kg bags of sugar does Sarah need to buy so that she can make the jam?

....

Q7 David is throwing a party for himself and 15 of his friends. He decides that it would be nice to make a bowl of fruit punch and carefully follows a recipe for 8 litres worth.

a) Will David fit all the punch in his 2 gallon bowl?

....

Read the questions carefully — it's ml for part b), then back to litres for part c).

b) Is there enough punch for everyone at the party to have one 500 ml glass?

....

c) In the end, 12 people split the punch equally between them. How many litres did they each drink?

....

Q8 The average fuel efficiency of two cars is shown below. Which car is the most efficient?

Car A — 53 miles per gallon

Car B — 4.5 litres per 100 km

Hint: 4.5 litres = 1 gallon

....

....

Estimating

Q1 Estimate the answers to these questions...

For example: 12 × 21 <u>10 × 20</u> = <u>200</u>

a) 18 × 12 .... × .... = ...

b) 23 × 21 .... × .... = ...

c) 57 × 46 .... × .... = ...

d) 98 × 145 .... × .... = ...

e) 11 ÷ 4 .... ÷ .... = ...

f) 22 ÷ 6 .... ÷ .... = ...

g) 97 ÷ 9 .... ÷ .... = ...

h) 147 ÷ 14 .... ÷ .... = ...

i) 195 × 205 .... × .... = ...

j) 545 × 301 .... × .... = ...

k) 901 ÷ 33 .... ÷ .... = ...

l) 1207 ÷ 598 .... ÷ .... = ...

Q2 Write in the estimates that give the answer shown.

For example: 101 × 96 <u>100 × 100</u> = 10 000

a) 34 × 19 .... × .... = 600

b) 27 × 32 .... × .... = 900

c) 67 × 89 .... × .... = 6300

d) 99 × 9 .... × .... = 1000

e) 56 ÷ 11 .... ÷ .... = 6

f) 119 ÷ 17 .... ÷ .... = 6

g) 182 ÷ 62 .... ÷ .... = 3

h) 317 ÷ 81 .... ÷ .... = 4

Q3 Andy earns a salary of £24 108 each year.

a) Estimate how much Andy earns per month. £ ....

b) Andy is hoping to get a bonus this year. The bonus is calculated as £1017 plus 9.7% of each worker's salary. Estimate the amount of Andy's bonus.

£ ....

c) Andy pays 10.14% of his regular salary into a pension scheme. Estimate how much money he has left per year after making his pension payments.

£ ....

Round off to NICE EASY CONVENIENT NUMBERS, then use them to do the sum. Easy peasy.

Reading Scales and Estimating

Q1 Read the following scales.
You'll need to add units to any answers which don't have them.

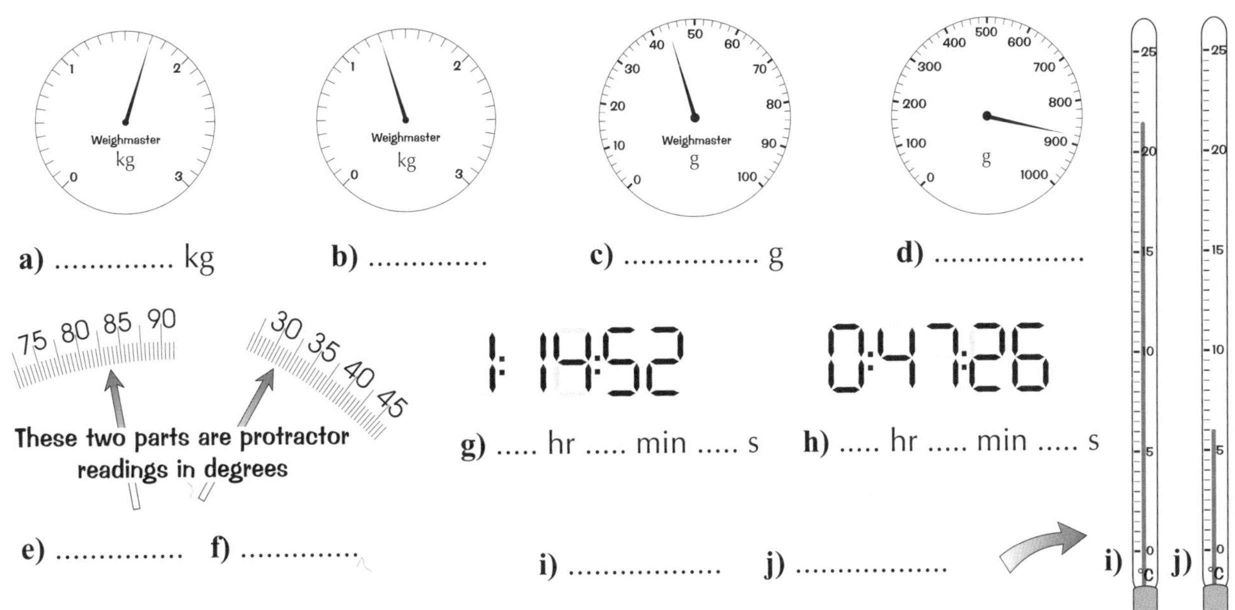

a) .... kg **b)** .... **c)** .... g **d)** ....

e) .... **f)** ....

g) ..... hr ..... min ..... s **h)** ..... hr ..... min ..... s

i) .... **j)** ....

Q2 Estimate the following lengths then measure them to see how far out you were:

OBJECT	ESTIMATE	ACTUAL LENGTH
a) Length of your pen or pencil	....	....
b) Width of your thumbnail	....	....
c) Height of this page	....	....
d) Length of the room you are in	....	....

Q3

The ranger is almost 2 m tall.
Estimate the height of the giraffe in metres.

....

If you have trouble estimating the height by eye, try measuring the ranger against your finger. Then see how many times that bit of finger fits into the height of the giraffe.

Drawing Angles

These instruments are used to measure angles.

An angle measurer.

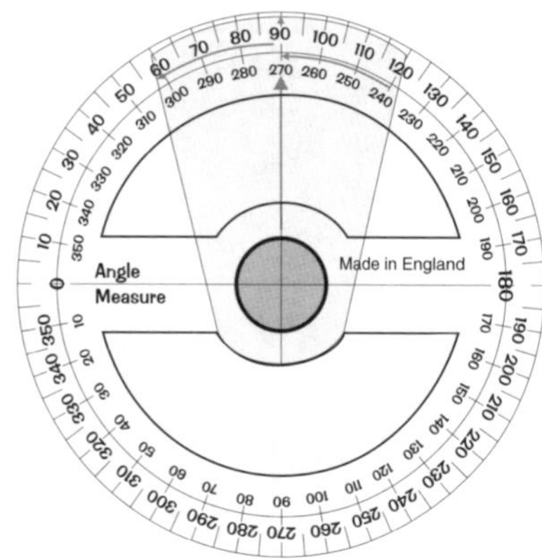

A protractor.

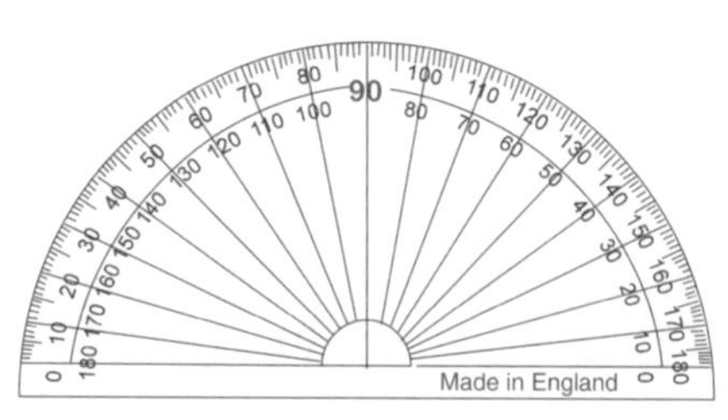

Don't forget these both have two scales — one going one way and one the other... so make sure you measure from the one that starts with 0°.

Q1 Use an angle measurer or protractor to help you to draw the following angles.

a) 20° **b)** 65° **c)** 90°

d) 136° **e)** 225° **f)** 340°

Q2 a) Draw an acute angle and measure it.

Acute angle measures ....°

b) Draw an obtuse angle and measure it.

Obtuse angle measures ....°

c) Draw a reflex angle and measure it.

Reflex angle measures ....°

Compass Directions

Q1

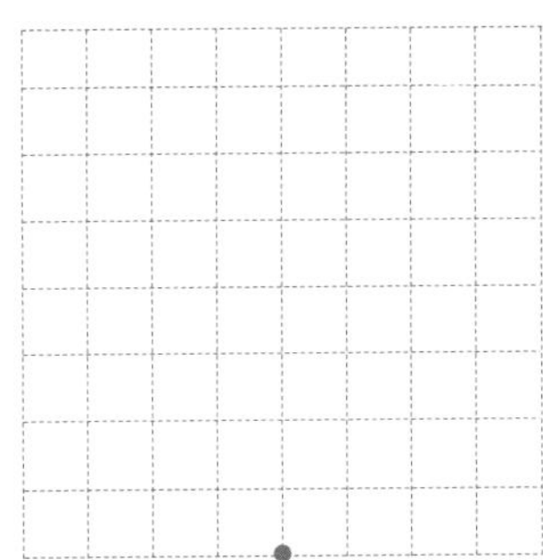

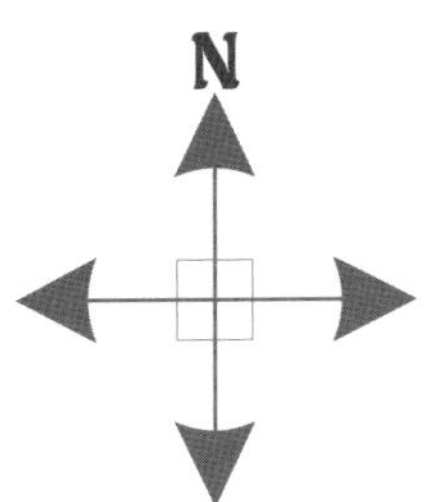

Start at the dot in the middle of the bottom line and follow the directions.

a) West 4 squares.
b) North 4 squares.
c) East 4 squares.
d) South 4 squares.
e) North East through 2 squares.
f) North 4 squares.
g) South West through 2 squares.
h) West 4 squares.
i) North East through 2 squares.
j) East 4 squares.

What shape have you drawn? ....

Q2

N

a) What direction does Jane go to get to Sue's house? ....

b) What direction is the church from Joe's house? ....

c) What is South East of Sue's house? ....

d) What is West of Sue's house? ....

e) Jane is at home. She is going to meet Sue in the park. They are going to the shop and then to Joe's house. Write down Jane's directions.

....

You could use "Naughty Elephants Squirt Water" but it's more fun to make one up — like Not Everyone Squeezes Wombats, or Nine Elves Storm Wales... (hmm)

Three Figure Bearings

Bearings always have three digits — even the small ones... in other words, if you've measured 60°, you've got to write it as 060°.

This is a map of part of a coastline. The scale is one cm to one km.

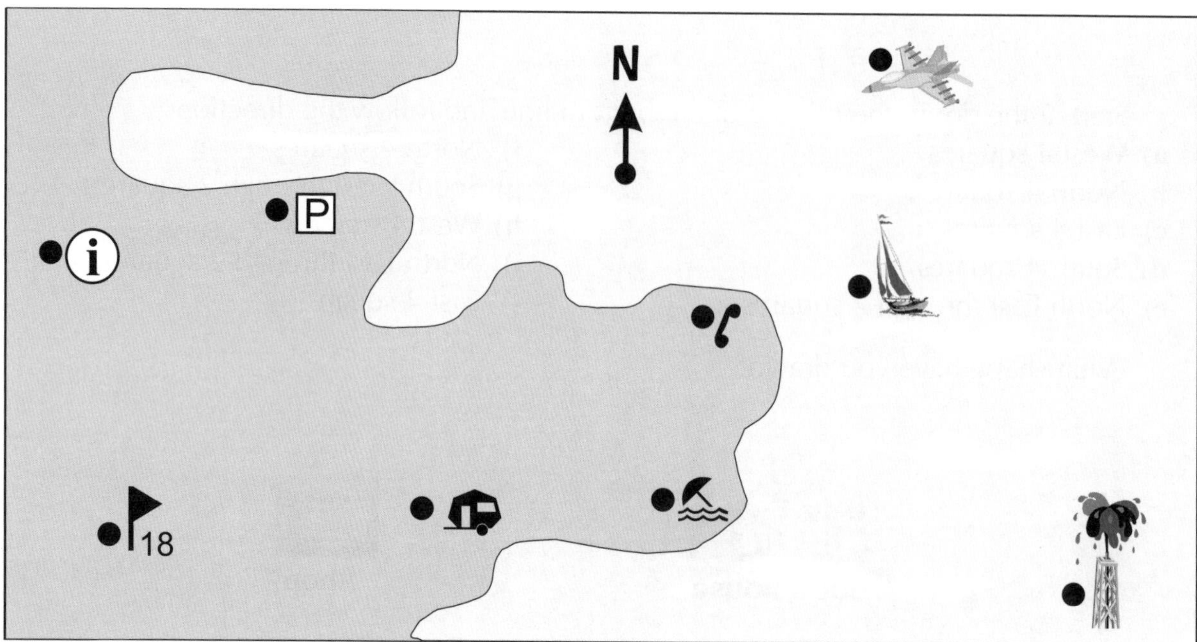

Q1 What is the bearing of the caravan from the (i)? ....

Q2 What is the bearing of the P from the water ski centre? ....

Q3 How far and on what bearing is:

a) The boat from the plane? ....

b) The boat from the oil rig? ....

c) The plane from the oil rig? ....

Q4 The water ski centre is on a bearing of 050° from the golf course, 18, and at a distance of 4.5 km. Put a ☆ where the water ski centre is.

Do the direction bit first — and draw a straight line. Then measure the correct distance along it.

Q5 There is a lighthouse on the coast. It is at a bearing of 300° from the oil rig and a bearing of 270° from the boat. Mark its position with a △.

Three Figure Bearings

Q6 This is a map of the Channel Islands.

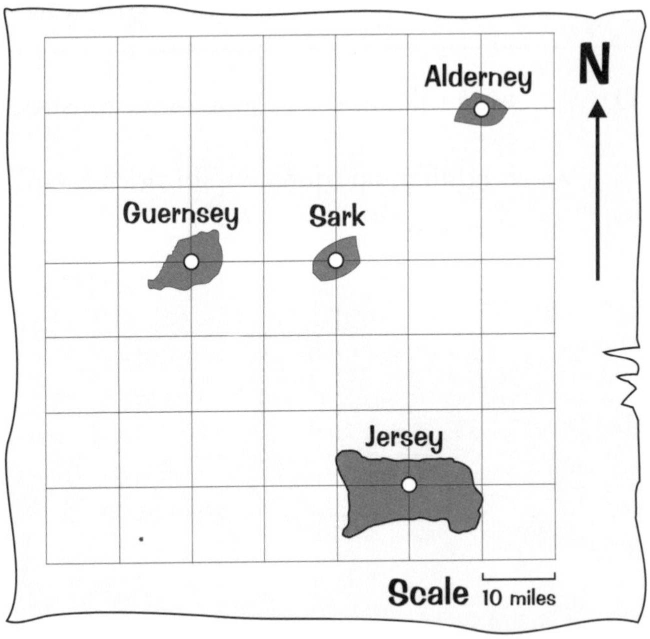

Alderney
N
Guernsey
Sark
Jersey
Scale 10 miles

a) Which island is furthest West?

....

b) Which island is due East of Guernsey?

....

The dots show the airports.

c) What bearing is needed to fly from Jersey to Guernsey? How far is it?

....

....

The flight from Jersey to Alderney goes directly over Sark Airport.

You need to look at the scale to work out distances.

d) What is the bearing for the first leg of the journey?

....

e) What is the bearing for the second leg of the journey?

....

f) Calculate the total distance flown from Jersey to Alderney.

....

Q7 Mr Brown is standing on the riverbank watching a cricket match on the other side. The bowler is on a bearing of 210° from Mr Brown and the batsman 190° from Mr Brown. When the batsman looks at the bowler he is looking along a bearing of 310°.

a) Draw a rough sketch to show this and put in all the angles given.

Start off by marking Mr Brown on the sketch, then measure the angles one by one.

b) Calculate the bearing of the batsman from the bowler.

....

c) Calculate the bearing of Mr Brown from the batsman.

....

Drawing Polygons

Drawing shapes is the most like Art that Maths will ever be, so make the most of it.

Q1 Use a ruler and a protractor to construct the following:

a) an equilateral triangle with sides 5 cm

b) a square with sides 6 cm

Q2 Construct these polygons in the space below:

a) a regular pentagon

b) a regular octagon

Start these by drawing an accurate circle with a pair of compasses.

Loci and Constructions

You've gotta be ultra neat with these — you'll lose easy marks if your pictures are scruffy — and let's face it, you'd rather have them, wouldn't you.

Constructions should always be done as accurately as possible using:

sharp pencil, ruler, compasses, protractor (set-square).

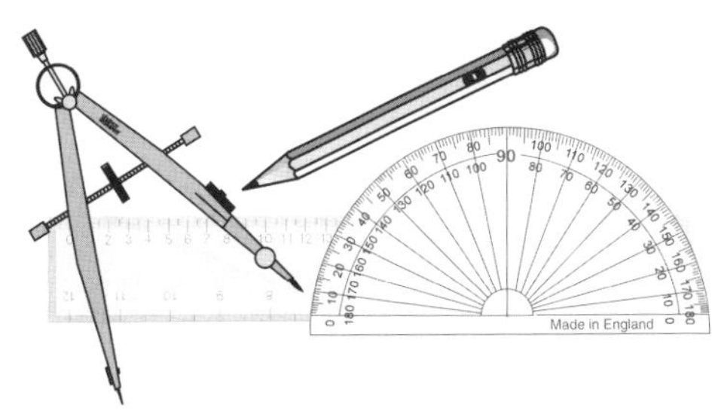

Q1 Draw a circle with radius 4 cm.

Draw in a diameter of the circle. Label one end of the diameter X and the other end Y.

Mark a point somewhere on the circumference — not too close to X or Y. Label your point T. Join X to T and T to Y.

Measure angle XTY.

Angle XTY = ...°

Q2 Construct the perpendicular bisector of AB.

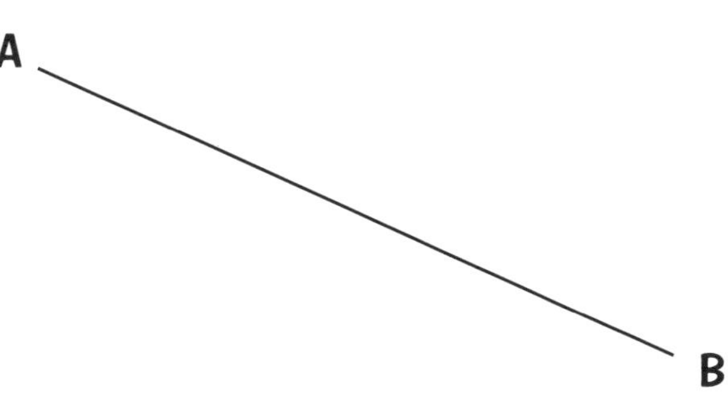

Loci and Constructions

Work through these questions bit by bit, and remember the following...

LOCUS — a line showing all points obeying the given rule.
BISECTOR — a line splitting an angle or line exactly in two.

Q3

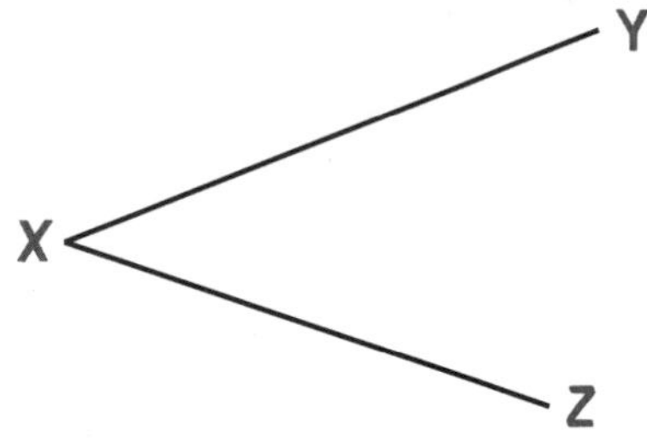

The diagram on the left shows two lines XY and XZ which meet at the point X. Construct the angle bisector of YXZ.

Q4 On a plain piece of paper mark two points A and B which are 6 cm apart.

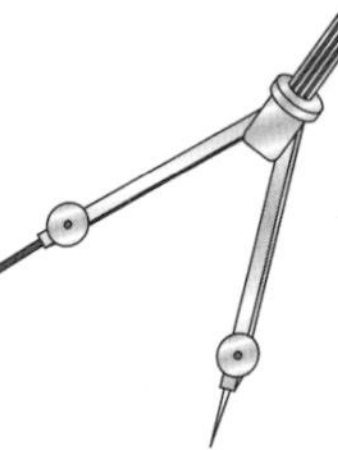

a) Draw the locus of points which are 4 cm from A.

b) Draw the locus of points which are 3 cm from B.

c) There are 2 points which are both 4 cm from A and 3 cm from B. Label them X and Y.

Q5 A planner is trying to decide where to build a house. He is using the plan on the right, which shows the location of a canal, a TV mast **(A)** and a water main **(B)**.

The house must be more than 100 m away from both the canal and the TV mast and be within 60 m of the water main.

Shade the area on the map where the planner could build the house.

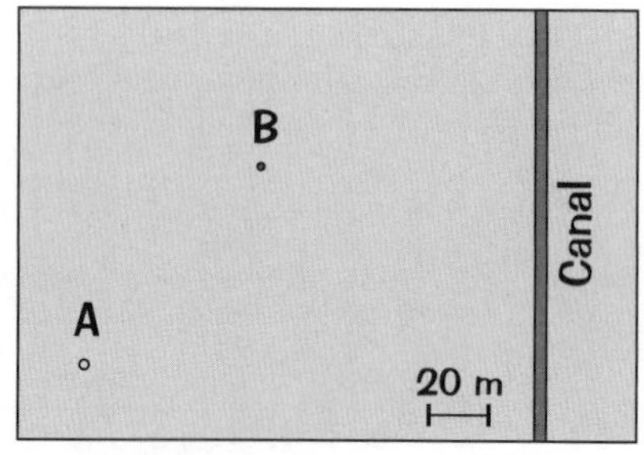